Multidimensional Queueing Models in Telecommunication Networks

Agassi Melikov • Leonid Ponomarenko

Multidimensional Queueing Models in Telecommunication Networks

Springer

Agassi Melikov
Department of Teletraffic Theory
Institute of Cybernetics, National Academy of Sciences of Azerbaijan
Baku
Azerbaijan

Leonid Ponomarenko
International Research and Training Center for Information Technologies and Systems of the National Academy of Sciences of Ukraine and Ministry of Education and Science of Ukraine
Kiev
Ukraine

ISBN 978-3-319-35706-5 ISBN 978-3-319-08669-9 (eBook)
DOI 10.1007/978-3-319-08669-9
Springer Cham Heidelberg New York Dordrecht London

Softcover reprint of the hardcover 1st edition 2014

Printed on acid-free paper

Springer is part of Springer Science+Business Media (www.springer.com)

Preface

The increasing complexity of telecommunication networks puts at the forefront the problem of developing adequate mathematical models for them. The main goals are finding their characteristics, solving the problems of their optimization subject to chosen criteria, and developing the corresponding control algorithms.

The basic mathematical tool that allows us to build both adequate analytical and numerical models of telecommunication networks is queueing theory. The core of this theory was founded more than 100 years ago in the pioneering work of Agner Erlang. He studied only the then recently telephony systems, but since then the models and methods of queueing theory have been widely used for studying service processes in various branches of science and industry, among them economics, manufacturing systems, military science, and transportation.

Remarkably, the most important reason for studying queueing theory, now as well as 100 years ago, is telecommunication networks. However, there are many distinctions between the models of the past and modern telecommunication networks. We should first note that in classical Erlang's models, it was assumed that calls do not differ from each other, i.e., calls are identical. In other words, early telecommunication networks were queueing systems with single traffic. However, in modern telecommunication networks, the calls (messages) essentially differ from each other with respect to some parameters. For example, calls can vary in arrival intensity and/or processing time, in the level of priorities, in the service mechanism, etc.

These facts show that classical queueing models with single traffic can serve only as rough (approximate) mathematical models of modern telecommunication networks. The functioning of modern telecommunication networks can be described only by means of queueing models with several types of traffic—adequate models of modern telecommunication networks are multidimensional ones. Such kinds of models are especially useful for studying integrated networks in which real-time calls, for voice, video, etc., and non-real-time calls, for data, fax, e-mail, etc., are handled.

Queueing models with single traffic are well studied, and they are described in well-known textbooks and monographs, but multidimensional queueing models are insufficiently studied, and there are only a few monographs on this theme. This book is devoted to the problem of applying multidimensional Markov models in modern telecommunication networks.

Unlike one-dimensional models, using exact and simple formulas to calculate the quality-of-service (QoS) metrics of multidimensional models is usually impossible. This is explained by the fact that in many cases the appropriate system of global balance equations (SGBE) for the steady-state probabilities has no explicit solutions—e.g., the solution in the multiplicative form. In such cases, various numerical (exact or approximate) methods must be used.

The classical approach to calculating the steady-state probabilities is based on the theory of multidimensional generating functions. However, there are well-known associated computational difficulties because we must solve systems of partial differential equations and equations for boundary states as well.

The alternative and more effective approach based on the use of SGBE for calculating the QoS metrics of multidimensional Markov models contains the following stages. First of all, note that this approach is used mainly for models with finite dimension of state space.

In the first stage, the state of the system is defined, and the set of all possible states (state space) is formed. As a rule, the system's state is described by a vector of corresponding dimension. In the second stage, an infinitesimal matrix (Q-matrix) of the appropriate multidimensional Markov chain (MC) is constructed. It is known that constructing the Q-matrix is enough to develop the SGBE. In the third stage, steady-state probabilities are found from the SGBE. In the final stage, the desired QoS metrics are calculated via steady-state probabilities, i.e., the QoS metrics are determined as appropriate marginal distributions of the initial multidimensional MC. By taking into account the unique property of Markov models, it is possible that the stationary probability of a state represents part of the sojourn time of the system in a corresponding state for a large supervision time interval.

The main problem in this approach is solving the SGBE, i.e., realizing the third stage, since the growing traffic and the increasing number of channels, as well as the buffer sizes of the corresponding telecommunication network, rapidly lead to an increase in the dimension of the state space. In some cases, by using the specific structure of a corresponding Q-matrix, it is possible to simplify this problem. So, for instance, it becomes simpler for networks that are described by models of reversible Markov chains since for such models the analytical solution in multiplicative form can be obtained. If the analytical solution of the SGBE is not available, as mentioned above, then various numerical methods are used. In this book, we use both approaches to investigate models of telecommunication networks.

The book consists of five chapters. In Chap. 1, both single-rate and multi-rate Erlang's models are considered, and we examine known computational algorithms to calculate their QoS metrics. Here also we propose hybrid access schemes in mono-service cellular networks without buffers, and we develop both exact and approximate methods to calculate the QoS metrics of such access schemes. In Chap. 2, we develop an analytical method to investigate the models of multi-rate

Erlang's models with randomized access schemes. We also consider analytical methods to study the models of multiservice cellular networks with various partition schemes of common radio channels. In Chap. 3, we investigate models of mono-service cellular networks. Two types of models are considered: models with either finite or infinite buffers for both types of new and handover calls and retrial models. Numerical algorithms to calculate their QoS metrics are developed. We investigate models of multiservice cellular networks with buffers in Chap. 4. Finally, in Chap. 5, we examine models of packet-switching networks with priorities. Here we examine in detail nonclassical priority schemes with multiple space and time priorities as well as jump priorities.

Each chapter of the book contains results from numerical experiments carried out using the developed algorithms, and each chapter contains comments and references that allow the reader to understand the current situation in the corresponding areas of research.

This book is recommended for researchers engaged with the mathematical theory of teletraffic. It will be useful for graduate and PhD students in informatics and applied mathematics as well as other fields.

Baku, Azerbaijan Agassi Melikov
Kiev, Ukraine Leonid Ponomarenko

List of Acronyms and Abbreviations

AV	Approximate value
BDP	Birth death process
BS	Base station
CAC	Call admission control
CLP	Cell loss probability
CP	Complete partition
CS	Complete sharing
CTD	Cell transfer delay
C_u	Channel utilization
$E_B(\nu, m)$	Erlang's B-formula for the model $M/M/m/0$ with load of ν erl
e_i	Unit vector in direction i in a Euclidean space whose dimension is specified in each particular case
EV	Exact value
FCFS	First come first served
GC	Guard channels
GM	Generating matrix
GPO	Generalized push-out
h-call	Handover call
H-cell	High priority cell
hd-call	Handover data call
hv-call	Handover voice call
$I(A)$	Indicator function of the event A
i-call	Call of type i or call from the traffic of type i
Int(x)	Integer (or whole) part of x
JP	Jump priorities
L-cell	Low priority cell
MC	Markov chain
MP	Multiple priorities
MRQ	Multi rate queue

MU	Mobile user
NPO	Non-push-out
NRT	Non-real-time
o-call	Original (new) call
od-call	Original (new) data call
ov-call	Original voice call
$P(A)$	Probability of the event A
$p(x)$, $\pi(x)$, $\rho(x)$	Stationary probability of state x
PB	Probability of blocking
PB_i	Probability of blocking of calls of type i
P_h	Probability of blocking (or dropping) of h-calls
P_o	Probability of blocking of o-calls
PO	Push-out
PS	Partial sharing
PSN	Packet switching network
$q(x, y)$	Transition intensity from state x to state y
QoS	Quality of service
RT	Real-time
SEV	Set of efficient values
SGBE	System of global balance equations
SLBE	System of local balance equations
SP	Space priorities
SPO	Simple push-out
TP	Time priorities
VP	Virtual partitioning
$x \to y$	Transition from state x to state y
$x^+ = \max(0, x)$	Choice a maximum 0 and x
$\lambda_h(\mu_h)$	Arrival (service) rate of handover calls
$\lambda_i(\mu_i)$	Arrival (service) rate of type i calls
$\lambda_o(\mu_o)$	Arrival (service) rate of new calls
$\nu_h = \lambda_h/\mu_h$	Load offered by h-calls
$\nu_i = \lambda_i/\mu_i$	Load offered by calls of type i
$\nu_o = \lambda_o/\mu_o$	Load offered by o-calls
$\delta(i,j) = \begin{cases} 1 & \text{if } i = j, \\ 0 & \text{if } i \neq j, \end{cases}$	Kronecker's symbols
$\varnothing$	Empty set
$\approx$	Approximate equality

Contents

Chapter 1
Multidimensional Loss Models of Queueing Systems and Their Applications in Telecommunication Networks

This chapter presents the foundations of multidimensional Erlang's models, and the discussion has been concentrated on Markov models with pure losses. Both single-rate and multi-rate Erlang's loss models with uncontrolled access schemes are considered. Effective computational procedures to calculate the main quality of service (QoS) metrics are shown.

A separate section is devoted to applications of multidimensional loss models with parametric access schemes in telecommunication networks. For simplicity here models of isolated cell of mono-service cellular networks with single-traffic class (second-generation networks) are investigated. Along with well-known access schemes (guard channels and cutoff schemes), a new hybrid access scheme is also considered. Both exact and approximate methods to calculate the QoS metrics are developed, and high accuracy of approximate solutions is shown. Problems related to selecting appropriate values of parameters of the proposed hybrid access scheme are solved.

1.1 Multi-flow Loss Models with Deterministic Access Schemes

Among multidimensional models of telecommunication networks, the most important is generalized version of classical single-traffic Erlang's model $M|M|N|0$. Here we consider multi-traffic Erlang's models with deterministic and uncontrolled access schemes.

First we consider single-rate models. The brief description of this system consists in the following. The unbuffered system contains N, $N > 1$, identical and parallel channels (slots, basic bandwidth units, etc., depending on specific technology). The call arrival process is a composition of K, $K > 1$, type Poisson flows with

A. Melikov, L. Ponomarenko, *Multidimensional Queueing Models in Telecommunication Networks*, DOI 10.1007/978-3-319-08669-9_1

intensities λ_i, $i = 1, 2, \ldots, K$. The service time for calls of type i (i-calls) has exponential distribution with means μ_i^{-1}, $i = 1, 2, \ldots, K$.

Uncontrolled call access scheme or call admission control (CAC) scheme is defined as follows. If at the arrival epoch of the call of any type in system there is at least one free channel, this call is accepted on service; otherwise an arrived call is lost. Usually such scheme is called complete sharing (CS) one.

Consider the problem of calculation the steady-state probabilities of this model. By taking into account the form of distribution functions both of arrival and service processes, we conclude that operating of the investigated queueing system might be described by K-dimensional Markov chain (MC). So, states of the system at equilibrium at any time are described by vectors $\boldsymbol{n} = (n_1, \ldots, n_K)$, where n_i is the number of i-calls in the system. Then state space (i.e., set of all possible states) S is defined as

$$S = \left\{ \boldsymbol{n} : n_i = 0, 1, \ldots, N; \sum_{i=1}^{K} n_i \leq N \right\}. \tag{1.1}$$

Transition intensities between states (i.e., elements of infinitesimal or Q-matrix) of the given K-dimensional MC are calculated as follows:

$$q(\boldsymbol{n}, \boldsymbol{n}') = \begin{cases} \lambda_i, & \text{if } \boldsymbol{n}' = \boldsymbol{n} + \boldsymbol{e}_i, \\ n_i \mu_i, & \text{if } \boldsymbol{n}' = \boldsymbol{n} - \boldsymbol{e}_i, \\ 0 & \text{in other cases,} \end{cases} \tag{1.2}$$

where $\boldsymbol{e}_i$ is the ith orthogonal vector in K-dimensional Euclidean space, $i = 1, 2, \ldots, K$.

Note 1.1 In special case when $\mu_i = \mu$ for all $i = 1, \ldots, K$, the state of the system might be described by scalar parameter k, which indicate the total number of calls in the system, $k = 0, 1, \ldots, N$. So, in this case the model of this system is one-dimensional birth-death process (1-D BDP) with parameters

$$q(k, k') = \begin{cases} \Lambda & \text{if } k' = k + 1, \\ k\mu, & \text{if } k' = k - 1, \\ 0 & \text{in other cases,} \end{cases}$$

where $\Lambda = \sum_{i=1}^{K} \lambda_i$.

Let $p(\boldsymbol{n})$ denote the probability of state $\boldsymbol{n} \in S$. Relation (1.2) allows to construct the system of global balance equations (SGBE) for the state probabilities:

$$\left(\sum_{i=1}^{K}\lambda_i I\left(N-\sum_{j=1}^{K}n_j>0\right)+\sum_{i=1}^{K}n_i\mu_i\right)p(\boldsymbol{n})=\sum_{i=1}^{K}\lambda_i p(\boldsymbol{n}-\boldsymbol{e}_i)I(n_i>0)+ \\ +\sum_{i=1}^{K}n_i\mu_i p(\boldsymbol{n}+\boldsymbol{e}_i)I\left(N-\sum_{j-1}^{K}n_j>1\right),\quad \boldsymbol{n}\in S; \tag{1.3}$$

$$\sum_{\boldsymbol{n}\in S}p(\boldsymbol{n})=1. \tag{1.4}$$

Henceforth, let $I(A)$ denote the indicator function of event A. Equation (1.4) is called normalizing condition over state space (1.1).

The solution of the SGBE (1.3), (1.4) has a multiplicative form:

$$p(\boldsymbol{n})=G^{(-1)}(N,K)\prod_{i=1}^{K}\frac{v_i^{n_i}}{n_{i!}},\quad \boldsymbol{n}\in S, \tag{1.5}$$

where $v_i=\lambda_i/\mu_i$, $G(N,K)$ is normalizing constant which provides condition (1.4), i.e.,

$$G(N,K)=\sum_{\boldsymbol{n}\in S}\prod_{i=1}^{K}\frac{v_i^{n_i}}{n_{i!}}. \tag{1.6}$$

From Eq. (1.5) we obtain $p(\boldsymbol{0})=G^{-1}(N,K)$, where $\boldsymbol{0}$ is K-dimensional zero vector.

To prove presentation (1.5) construct the system of local balance equations (SLBE). The SLBE has the following form:

$$q(\boldsymbol{n},\boldsymbol{n}')p(\boldsymbol{n})=q(\boldsymbol{n}',\boldsymbol{n})p(\boldsymbol{n}'),\ \forall\ \boldsymbol{n},\boldsymbol{n}'\in S,$$

or in explicit form

$$\lambda_i p(\boldsymbol{n})=(n_i+1)\mu_i p(\boldsymbol{n}+\boldsymbol{e}_i),\forall\ \boldsymbol{n},\ \boldsymbol{n}+\boldsymbol{e}_i\in S. \tag{1.7}$$

Now show that Eq. (1.5) is the solution of SLBE (1.7). Indeed, from Eqs. (1.5) and (1.7) we have

$$\frac{p(\boldsymbol{n}+\boldsymbol{e}_i)}{p(\boldsymbol{n})}=\frac{v_i}{n_i+1}=\frac{\lambda_i}{(n_i+1)\mu_i},$$

or in other words, Eq. (1.5) is the solution of SLGE (1.7); thus Eq. (1.5) is the solution of SGBE (1.3), (1.4).

The main characteristic of the given model is stationary probability of blocking (PB) By using the PASTA theorem [36] we conclude that this characteristic is same for each traffic class, i.e.,

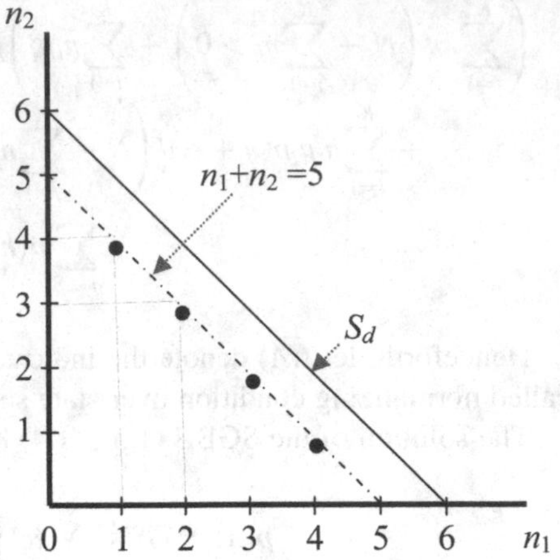

Fig. 1.1 State space (1.1) for $N=6$, $K=2$

$$\mathrm{PB} = \sum_{\boldsymbol{n} \in S_\mathrm{d}} p(\boldsymbol{n}), \tag{1.8}$$

where $S_\mathrm{d} = \{\boldsymbol{n} \in S : \sum_{i=1}^{K} n_i = N\}$ is set of diagonal states (see Fig. 1.1).

It is more important to note that multiplicative solution (1.5) is independent on the form of the service time distribution function with fixed mean value. Moreover it is true for closed models, i.e., for the models with finite number of sources.

Calculation of the steady-state probabilities by means of Eq. (1.5) is connected with huge difficulties for models with many channels and types of traffic. This results from the fact that direct calculation $G(N,K)$ via the formula (1.6) is accompanied by overflow (at $v_i > 1$) and order disappearance (at $v_i \to 0$).

To overcome these specified difficulties, various approaches are offered. One of the effective approaches is the use of Buzen's algorithm [4] based on two-dimensional recurrent formulas. In this algorithm normalizing constant is calculated as

$$G(j,i) = \sum_{l=0}^{j} \frac{v_i^l}{l!} G(j-l, i-1), \quad i = 2, \ldots, K, \quad j = 0, 1, \ldots, N;$$

$$G(j,1) = \sum_{l=0}^{j} \frac{v_1^l}{l!}. \tag{1.9}$$

An alternative method is based on principles of state space merging approach. In this case the following splitting of state space (1.1) is considered:

$$S = \cup_{j=0}^{N} S_j, S_i \cap S_j = \varnothing, \quad i \neq j, \tag{1.10}$$

where $S_j = \{\boldsymbol{n} \in S : \sum_{i=1}^{K} n_i = j\}$ contains all microstates from Eq. (1.1) in which the total number of calls equals j, $j = 0, 1, \ldots, N$ (see Fig. 1.1).

Stationary probability of merged state S_j is denoted by $\pi(j)$, $j = 0, 1, \ldots, N$. These probabilities are defined as

$$\pi(j) = \sum_{\boldsymbol{n} \in S_j} \prod_{i=1}^{K} \frac{v_i^{n_i}}{i!} G^{-1}(N, K), \quad j = 0, 1, \ldots, N. \tag{1.11}$$

Proposition 1.1 Probabilities of merged states satisfy the following system of equations:

$$v\pi(j-1) = j\pi(j), \quad j = 0, 1, \ldots, N;$$
$$\sum_{j=0}^{N} \pi(j) = 1, \tag{1.12}$$

where $v = \sum_{i=1}^{K} v_i$, $\pi(j) = 0$ if $x < 0$.

For the proof of this fact we will preliminary prove the following lemma.

Lemma 1.1

$$v_i \pi(j-1) = E\big(n_i \big| j\big)\pi(j), \quad i = 1, 2, \ldots, K; \quad j = 0, 1, \ldots, N, \tag{1.13}$$

where $E(\cdot|\cdot)$ is symbol of conditional mathematical expectation.

Proof of Lemma 1.1 Rewrite the SLBE (1.7) in the following form:

$$v_i \gamma_i(\boldsymbol{n}) p(\boldsymbol{n} - \boldsymbol{e}_i) = n_i p(\boldsymbol{n}), \quad i = 1, 2, \ldots, K, \tag{1.14}$$

where $\gamma_i(\boldsymbol{n}) = \begin{cases} 1, & \text{if } n_i \geq 1, \\ 0, & \text{if } n_i = 0. \end{cases}$

Summing Eq. (1.14) over all $\boldsymbol{n} \in S_j$, we have

$$v_i \sum_{\boldsymbol{n} \in S_i} \gamma_i(\boldsymbol{n}) p(\boldsymbol{n} - \boldsymbol{e}_i) = \sum_{\boldsymbol{n} \in S_i} n_i p(\boldsymbol{n}). \tag{1.15}$$

Transform the left side of Eq. (1.15) as follows:

$$v_i \sum_{\boldsymbol{n} \in S_i} \gamma_i(\boldsymbol{n}) p(\boldsymbol{n} - \boldsymbol{e}_i) = v_i \sum_{\boldsymbol{n} \in S_j \cap \{n_i \geq 1\}} p(\boldsymbol{n} - \boldsymbol{e}_i), \tag{1.16}$$

where $S_j \cap \{n_i \geq 1\} = \{\boldsymbol{n} \in S_j : \sum_{l \neq i} n_l + (n_i - 1) = j - 1, \quad n_l \geq 0, \quad l \neq i\}$.

So we have

$$v_i \sum_{\boldsymbol{n} \in S_j} \gamma_i(\boldsymbol{n}) p(\boldsymbol{n} - \boldsymbol{e}_i) = v_i \pi(j-1), \quad j \geq 1. \tag{1.17}$$

Now transform the right side of Eq. (1.15) as follows:

$$\sum_{\boldsymbol{n} \in S_i} n_i p(\boldsymbol{n}) = \sum_{\boldsymbol{n} \in S_i} n_i \frac{p(\boldsymbol{n})}{\pi(j)} \pi(j), \quad j = 0, 1, \ldots, N, \quad i = 1, 2, \ldots, K.$$

Since

$$P(\boldsymbol{n}|j) = \begin{cases} \dfrac{p(\boldsymbol{n})}{\pi(j)}, & \text{if } \boldsymbol{n} \in S_j, \\ 0 & \text{otherwise,} \end{cases}$$

we have

$$\sum_{n \in S_i} n_i p(\boldsymbol{n}) = \left(\sum_{n \in S_i} n_i P(\boldsymbol{n}|j) \right) \pi(j) = E(n_i|j)\pi(j), \forall j = 0, 1, \ldots, N, \ i = 1, 2, \ldots, K. \tag{1.18}$$

Thus, from Eqs. (1.17) and (1.18), we conclude that

$$v_i \pi(j-1) = E(n_i|j)\pi(j), \quad j = 0, 1, \ldots, N, \quad i = 1, 2, \ldots, K. \tag{1.19}$$

Now to prove proposition 1.1, we take summation in both sides of Eq. (1.19) for all $i = 1, \ldots, K$. Then we have

$$v\pi(j-1) = \left(\sum_{i=1}^{K} E(n_i|j) \right) \pi(j) = E\left(\sum_{i=1}^{K} n_i|j \right) \pi(j) = j\pi(j).$$

The algorithm (1.12) has some advantages with comparison with Buzen's algorithm. First, unlike the Buzen's algorithm equations (1.12) are one-dimensional recurrent equations for arbitrary K, and second the last algorithm does not depend on K.

By using Eqs. (1.6) and (1.11) from Eq. (1.12), we obtain

$$G^{-1}(N, K) = \pi(0)\ , \text{PB} = \pi(N).$$

The algorithm (1.12) is the generalized version of the well-known Erlang's formulas for multi-flow systems:

$$\pi(j) = \frac{\frac{\nu^j}{j!}}{\sum_{i=0}^{N} \frac{\nu^i}{i!}}, \quad j = 0, 1, \ldots, N.$$

Described above computational procedures can be trivially generalized to multi-rate queues (MRQ) with CAC based on CS scheme. In MRQ, i-call requests $b_i \geq 1$ channels whose service start and end times are simultaneous; i-call is blocked and lost if at its arrival moment the number of free channels is less than b_i, $i = 1, 2, \ldots, K$.

For MRQ state space is defined as

$$S = \left\{ \boldsymbol{n} : n_i = 0, 1, \ldots, \left[\frac{N}{b_i}\right], \ \boldsymbol{n} \cdot \boldsymbol{b} \leq N \right\}, \tag{1.20}$$

where $[x]$ denotes the integer part of x, $\boldsymbol{b} = (b_1, \ldots, b_K)$, $\boldsymbol{n} \cdot \boldsymbol{b} = \sum_{i=1}^{K} n_i b_i$.

Note 1.2 Unlike the classical single-rate models, in multi-rate models even in special case when $\mu_i = \mu$ for all $i = 1, \ldots, K$, it is impossible to describe the state of system by scalar parameter k, which denotes the total number of busy channels, $k = 0, 1, \ldots, N$.

Note 1.3 Hereinafter, for simplicity, we use the same notations for state spaces, stationary distributions, etc., in different models. This should not lead to misunderstanding, as it will be clear what model is considered from the context.

Stationary distribution of this model has multiplicative form (1.5) also, but in this case normalizing constant $G(N, K)$ is defined over state space (1.20).

Unlike the classical multidimensional models, in the MRQ models, blocking probabilities of polytypic calls differ from each other. So, blocking probability of i-calls (PB_i) is calculated as (see Fig. 1.2)

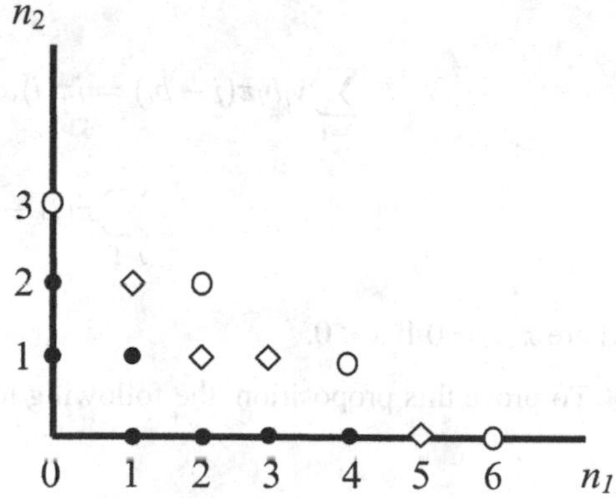

Fig. 1.2 State space (1.20) for $N = 6$, $K = 2$, $\boldsymbol{b} = (1, 2)$; *open circles*, blocking of calls of both type; *open diamonds*, blocking of calls of type 2 only

$$\mathrm{PB}_i = \sum_{n \in S} p(\boldsymbol{n}) I(f(\boldsymbol{n}) < b_i), \tag{1.21}$$

where $f(\boldsymbol{n}) = N - (\boldsymbol{n} \cdot \boldsymbol{b})$ denotes the number of free channels in state $\boldsymbol{n} \in S$

Other QoS metrics of the MRQ models is the channel utilization (C_{u}) which is measured via the average number of busy channels. This metric is calculated as follows:

$$C_{\mathrm{u}} = \sum_{n \in S} (\boldsymbol{n} \cdot \boldsymbol{b}) p(\boldsymbol{n}). \tag{1.22}$$

Buzen's algorithm to calculate the normalizing constant for MRQ models is defined as follows:

$$G(j,i) = \sum_{l=0}^{\left[\frac{j}{b_i}\right]} \frac{v_i^l}{l!} G(j - lb_i, i-1), \quad i = 2, \ldots, K, \quad j = 0, 1, \ldots, N,$$

$$G(j,1) = \sum_{l=0}^{\left[\frac{j}{b_i}\right]} \frac{v_1^l}{l!}.$$

The alternative way to calculate the steady-state probabilities of MRQ models is the Kaufman [17] and Roberts [31] algorithm.

The following splitting of state space (1.20) is considered:

$$S = \cup_{j=0}^{N} S_j, S_i \cap S_j = \varnothing, \quad i \neq j,$$

where $S_j = \{\boldsymbol{n} \in S : \boldsymbol{n} \cdot \boldsymbol{b} = j\}$, i.e., the class of states S_j, contains all microstates from Eq. (1.20) in which the number of busy channels equals j, $j = 0, 1, \ldots, N$.

Proposition 1.2 Probabilities of merged states in MRQ model satisfy the following system of equations:

$$\sum_{i=1}^{K} v_i b_i \pi(j - b_i) = j\pi(j), \quad j = 0, 1, \ldots, N;$$

$$\sum_{j=0}^{N} \pi(j) = 1, \tag{1.23}$$

where $\pi(x) = 0$ if $x < 0$.

To prove this proposition, the following lemma is used.

Lemma 1.2

$$v_i\pi(j-b_i)=E\left(n_i|j\right)\pi(j)\,,\quad i=1,2,\ldots,K;\quad j=0,1,\ldots,N.$$

The proof of lemma 1.2 and proposition 1.2 is made similarly to proofs of lemma 1.1 and proposition 1.1, respectively.

The blocking probabilities PB_i are calculated from the stationary distribution of merged model in the following way:

$$\mathrm{PB}_i=\sum_{j=0}^{b_i-1}\pi(N-j),\quad i=1,\ldots,K. \tag{1.24}$$

The following algorithm is modification of the Kaufman–Roberts algorithm for the models MRQ with CAC based on CS scheme.

Proposition 1.3 The QoS metrics (1.21) and (1.22) are calculated as follows:

$$\mathrm{PB}_i=\left(\sum_{j=N-r+1}^{N}g_j\right)\Big/\left(\sum_{j=0}^{N}g_j\right),\ i\in A(r), \tag{1.25}$$

$$C_{\mathrm{u}}=\left(\sum_{i=1}^{N}ig_i\right)\Big/\left(\sum_{i=0}^{N}g_i\right), \tag{1.26}$$

where $A(r)=\{i:\text{calls of type } i \text{ require } r \text{ channels}\},\ r=1,\ldots,N;$

$$g_j=\begin{cases}1, & \text{if } j=0,\\ \dfrac{1}{j}\displaystyle\sum_{i=1}^{j}i\alpha_i g_{j-i}, & \text{if } j=1,\ldots,N;\end{cases}$$

$$\alpha_r=\begin{cases}0, & \text{if } A(r)=\varnothing,\\ \displaystyle\sum_{i\in A(r)}v_i, & \text{if } A(r)\neq\varnothing.\end{cases}$$

To achieve the absolute fair handling in MRQ in terms of equalization of blocking probabilities of the heterogeneous calls, the CAC based on complete sharing with equalization (CSE) is proposed [6]. This scheme is defined as follows: the newly arrived call is accepted if at the moment of arrival the number of free channels is more than or equal to b, where $b=\max\{b_i:i=1,\ldots,K\}$.

Proposition 1.4 The blocking probability in MRQ with CAC based on CSE scheme is calculated as follows:

$$\mathrm{PB} = \left(\sum_{j=N-b+1}^{N} g_j\right) \Big/ \left(\sum_{j=0}^{N} g_j\right), \tag{1.27}$$

where

$$g_j = \begin{cases} 1, & \text{if } j = 0, \\ \dfrac{1}{j}\displaystyle\sum_{i=1}^{j} i\alpha_i g_{j-i} F_i(j-i), & \text{if } j = 1, \ldots, N; \end{cases}$$

$$F_j(i - b_j) = \begin{cases} 1 & \text{if } i - b_j \le N - b, \\ 0 & \text{otherwise.} \end{cases}$$

The generalization of two CAC based on CS and CSE schemes is CAC based on trunk reservation (TR) scheme [30]. This scheme is defined as follows: the newly arrived i-call is accepted if at the moment of arrival the number of free channels is more than or equal to $b_i + r_i$, where $0 \le r_i \le N - b_i$, $i = 1, \ldots, K$.

Proposition 1.5 The blocking probability in MRQ with CAC based on TR scheme is calculated as follows:

$$\mathrm{PB} = \left(\sum_{j=N-i-r_i+1}^{N} g_j\right) \Big/ \left(\sum_{j=0}^{N} g_j\right), \tag{1.28}$$

where

$$g_j = \begin{cases} 1, & \text{if } j = 0, \\ \dfrac{1}{j}\displaystyle\sum_{i=1}^{j} i\alpha_i g_{j-i} G_i(j-i), & \text{if } j = 1, \ldots, N; \end{cases}$$

$$G_i(j) = \begin{cases} 1 & \text{if } j \le N - b_i - r_i, \\ 0 & \text{otherwise.} \end{cases}$$

In both CAC based on CSE scheme and TR scheme, channel utilization is calculated like formula (1.26).

Proofs of propositions 1.3–1.5 might be found in [29], Chap. 4. Note that a major advantage of the last algorithms is their low computational complexity.

1.2 Applications of Multi-flow Loss Models in Mono-service Cellular Networks

Uncontrolled multidimensional queueing models have been above considered. At the same time, in practice several parameters of the models of telecommunication networks might be controlled in some way. In most cases, controllable parameters are call admission control ones.

In this section, applications of multidimensional queueing models in telecommunication networks are shown. The problem of finding the appropriate access scheme is an actual issue. It is crucial to guaranteeing maximum throughput and QoS fulfillment, and it is critical to define efficient algorithms to solve the indicated problem. In telecommunication networks, access scheme decides the amount of resources that must be assigned to each call, and it also defines the state-dependent rules to determine to either accept or reject the arrived calls in real-time regime.

Here for simplicity and concreteness, the models of mono-service cellular networks (e.g., traditional voice-oriented networks) are considered.

In cellular networks, when a subscriber crosses the boundary of a cell (while on a call), the subscriber releases this cell's channel and requests an empty channel in a neighboring cell. This process is called a handover. If a neighboring cell has at least one empty channel, then such a handover call (*h*-call) is delivered continuously and almost transparently to the subscriber; otherwise, the call is dropped. Usually, dropping an ongoing call from a different cell is less desirable than blocking new attempts originating within the cell (*o*-calls). Thus, *h*-calls are considered to be more important (have a higher priority) than *o*-calls.

Here we will analyze the model of a cell belonging to a homogeneous cellular network that experiences the same traffic patterns. Henceforth, a homogeneous network is defined as one for which the traffic parameters of all cells within the network are statistically identical; that is, we can study the operation of a representative cell in isolation. This assumption is true for the networks with small cells (e.g., networks with microcells).

Our models in this section are based on the following assumptions.

- The mentioned cell contains N channels, $1 < N < \infty$, which are intended to handle the Poisson flows of new and handover calls. The intensity of x-calls equals λ_x, $x \in \{o, h\}$.
- Channel occupancy times are defined not only by the required service time of different types of calls but also by the mobility of subscribers within cells. We assume that the required service times of x-calls and the duration for which the subscribers reside within the cell are exponentially distributed with different parameters τ_x and γ_x, $x \in \{o, h\}$, respectively. Thus, the channel occupancy time of x-calls is defined as the minimum of two exponentially distributed stochasticvariables, i.e., for x-calls, it has an exponential form with parameter $\mu_x = \tau_x + \gamma_x$, $x \in \{o, h\}$. In other words, the distribution functions of the channel

occupancy time of both types of calls are exponential, but their parameters are different: the intensity of handling new (handover) calls equals $\mu_o(\mu_h)$, and generally speaking, $\mu_o \neq \mu_h$.

1.2.1 Model with Guard Channels

The classical guard channel (GC) scheme is defined as follows. If at least one free channel exists upon the arrival of an *h*-call, then the call seizes one of the free channels; otherwise, the call is dropped. A new arriving call is accepted only when the number of busy channels is less than *g*, for some fixed *g* satisfying $1 \le g \le N$ (in the case where $g = N$, there are no restrictions on new calls); otherwise, the new call is blocked.

Cell functionality is described by a two-dimensional Markov chain (2-D MC); that is, the state of the given system at an arbitrary moment in time is described by a two-dimensional vector $\boldsymbol{k} = (k_o, k_h)$ where $k_o(k_h)$ indicates the number of *o*-calls (*h*-calls) in the cell. Thus, the state space of the corresponding 2-D MC is determined as follows:

$$S = \{\boldsymbol{k} : k_o = 0, 1, \ldots, g; \quad k_h = 0, 1, \ldots, N; \quad k_o + k_h \le N\}. \tag{1.29}$$

The nonnegative elements of Q-matrix of this 2-D MC are determined as follows:

$$q(\boldsymbol{k}, \boldsymbol{k}') = \begin{cases} \lambda_o, & \text{if } k_o + k_h < g,\ \boldsymbol{k}' = \boldsymbol{k} + \boldsymbol{e}_1, \\ \lambda_h, & \text{if } \boldsymbol{k}' = \boldsymbol{k} + \boldsymbol{e}_2, \\ k_o\mu_o, & \text{if } \boldsymbol{k}' = \boldsymbol{k} - \boldsymbol{e}_1, \\ k_h\mu_h, & \text{if } \boldsymbol{k}' = \boldsymbol{k} - \boldsymbol{e}_2, \\ 0 & \text{in other cases,} \end{cases} \tag{1.30}$$

where $\boldsymbol{e}_1 = (1, 0)$, $\boldsymbol{e}_2 = (0, 1)$.

Let us denote the stationary probability of state $\boldsymbol{k} \in S$ as $p(\boldsymbol{k})$. These probabilities are found from the appropriate SGBE which is developed by using Eq. (1.30):

$$\begin{aligned} &(\lambda_o I(k_o + k_h < g) + \lambda_h I(k_o + k_h < N) + k_o\mu_o + k_h\mu_h) p(\boldsymbol{k}) \\ &= \lambda_o p(\boldsymbol{k} - \boldsymbol{e}_1) I(k_o > 0) \ + \lambda_h p(\boldsymbol{k} - \boldsymbol{e}_2) I(k_h > 0) + \\ &+ (k_o + 1)\mu_o p(\boldsymbol{k} + \boldsymbol{e}_1) + (k_h + 1) p(\boldsymbol{k} + \boldsymbol{e}_2)\,. \end{aligned} \tag{1.31}$$

The normalizing condition has the following form:

$$\sum_{\boldsymbol{k} \in S} p(\boldsymbol{k}) = 1. \tag{1.32}$$

Then SGBE (1.31), (1.32) is used to calculate all the required QoS metrics of the model. The primary QoS metrics are the loss probability of h-calls (P_h) and the blocking probability of o-calls (P_o). They are defined as follows:

$$P_o = \sum_{k \in S} p(k) I(k_o + k_h \geq g), \tag{1.33}$$

$$P_h = \sum_{k \in S} p(k) \delta(k_o + k_h, N), \tag{1.34}$$

where $\delta(i,j)$ are Kronecker's symbols.

From formulas (1.33), (1.34), we obtain $P_o = P_h$ when $g = N$, i.e., the loss probability of h-calls and the blocking probability of o-calls are equal each other when CS scheme is used.

Other QoS metrics, namely channel utilization (C_u) which is measured by mean number of busy channels, are also calculated via steady-state probabilities as follows:

$$C_u = \sum_{i=1}^{N} i \sigma_i, \tag{1.35}$$

where $\sigma_i = \sum_{k \in S} p(k) \delta(k_o + k_h, i)$, $i = 1, 2, \ldots, N$, are marginal probability mass functions.

It is important to note that the given 2-D MC has a reversibility property [22] only in a special case $g = 0$ and hence for its steady-state probabilities has a multiplicative form (see Sect. 1.1). In other cases (i.e., when $g \neq 0$), this MC is not reversible. Indeed, according to relation (1.30), there exists the transition $(k_o, k_h) \rightarrow (k_o - 1, k_h)$ with intensity $k_o \mu_o$ where $k_o + k_h \geq g$, but the inverse transition does not exist.

The last means that for calculating the steady-state probabilities at concrete values of loading parameters of polytypic traffic and number of channels, it is necessary to solve the corresponding system of the linear algebraic equations. For models with moderate dimension (about some thousand states), this system of the linear equations is easily solved by means of existing software. For models with large dimension to overcome computational difficulties, known approximate approaches might be used [5].

1.2.2 Model with Cutoff Scheme

Now consider other schemes for assigning the high priority to h-calls. As in previous scheme, here h-call is accepted if upon its arrival moment there is at least one free channel. However, in considered scheme (cutoff scheme), decision either accepts or rejects new calls depending on the number of such kind of calls in a cell. In other words, a new arriving call is accepted only when the number of

o-calls is less than b, for some fixed b satisfying $1 \le b \le N$ (in the case where $b = N$, there are no restrictions on new calls); otherwise, the new call is blocked.

Cell functionality under such access scheme also is described by the 2-D MC with states $\boldsymbol{k} = (k_o, k_h)$ where $k_o(k_h)$ indicates the number of o-calls (h-calls) in the cell. In this scheme, the state space of the corresponding 2-D MC and the nonnegative elements of its generating matrix are determined like Eqs. (1.29) and (1.30), respectively, i.e.,

$$q(\boldsymbol{k},\boldsymbol{k}') = \begin{cases} \lambda_o, & \text{if } k_o < b,\ \boldsymbol{k}' = \boldsymbol{k} + \boldsymbol{e}_1, \\ \lambda_h, & \text{if } \boldsymbol{k}' = \boldsymbol{k} + \boldsymbol{e}_2, \\ k_o\mu_o, & \text{if } \boldsymbol{k}' = \boldsymbol{k} - \boldsymbol{e}_1, \\ k_h\mu_h, & \text{if } \boldsymbol{k}' = \boldsymbol{k} - \boldsymbol{e}_2, \\ 0 & \text{in other cases.} \end{cases}$$

In this scheme, the loss probability of h-calls and channel utilization are calculated like Eqs. (1.34) and (1.35), respectively. New calls are blocked if upon their arrivals moments either (a) all channels are busy or (b) number of o-calls is equal b regardless number of busy channels. Since in this scheme the blocking probability of o-calls is calculated as

$$P_o = \sum_{\boldsymbol{k} \in S} p(\boldsymbol{k})(\delta(k_o + k_h, N) + (1 - \delta(k_o + k_h, N))\delta(k_o, b)). \tag{1.36}$$

Steady-state probabilities satisfy the following SGBE:

$$\begin{aligned}(\lambda_o I(k_o < b) + \lambda_h I(k_o + k_h < N) + k_o\mu_o + k_h\mu_h)p(\boldsymbol{k}) &= \lambda_o p(\boldsymbol{k} - \boldsymbol{e}_1)I(k_o > 0) \\ &+ \lambda_h p(k - \boldsymbol{e}_2)I(k_h > 0) \\ &+ (k_o + 1)\mu_o p(\boldsymbol{k} + \boldsymbol{e}_1) + (k_h + 1)p(\boldsymbol{k} + \boldsymbol{e}_2)\,.\end{aligned} \tag{1.37}$$

Unlike the previous scheme, here the appropriate 2-D MC has a reversibility property for any possible values of b, and hence SGBE (1.37) together with appropriate normalizing condition has a multiplicative solution, i.e., steady-state probabilities are determined as

$$p(k_o, k_h) = G^{-1}(N, b)\frac{v_o^{k_o}}{k_o!}\frac{v_h^{k_h}}{k_h!}, \tag{1.38}$$

where $G^{-1}(N, b)$ is normalizing constant over appropriate state space.

Thus under cutoff scheme, QoS metrics of the cell are calculated by using explicit formulas.

1.2.3 Model with Hybrid Access Scheme

In the access scheme based on the classical guard channel one, h-calls almost completely occupy the channels, and therefore o-calls are very often lost. In the access scheme based on a cutoff one, it is impossible to effectively use channels' capacity. The defined below hybrid access scheme has two degrees of freedom and helps avoid such situations.

The proposed hybrid CAC scheme is defined as follows. If, upon the arrival of an h-call, there is at least one free channel and the number of such types calls in the channels is less than r, $1 \le r \le N$, then the arriving h-call is accepted (for $r = N$ there are no restrictions for h-calls); otherwise, the arriving h-call is dropped. An arriving o-call is accepted only when the number of busy channels is less than g, $1 \le g \le N$ (for $g = N$ there are no restrictions for o-calls). It is clear that the condition $r + g \ge N$ must be satisfied; otherwise, $N - g - r$ channels of the cell will be unused.

Special cases:

1. From the given access scheme, we obtain a complete sharing scheme when $g = r = N$.
2. From the given access scheme, we obtain a scheme based on a guard channel scheme when $r = N$.

Here we develop both exact and approximate methods to calculation of QoS metrics of the investigated hybrid access scheme.

First consider exact method. As in previous models, the state of the system at an arbitrary moment in time is described by a two-dimensional vector $\boldsymbol{k} = (k_o, k_h)$, where k_o (k_h) indicates the number of new (handover) calls in the channels. Thus, the state space of the corresponding 2-D MC is determined as follows:

$$S = \{\boldsymbol{k} : k_o = 0, 1, \ldots, g; \quad k_h = 0, 1, \ldots, r; \quad k_o + k_h \le N\}. \tag{1.39}$$

The nonnegative elements of the Q-matrix of this 2-D MC are defined as follows (see Fig. 1.3):

$$q(\boldsymbol{k}, \boldsymbol{k}') = \begin{cases} \lambda_o, & \text{if } k_o + k_h < g, \ \boldsymbol{k}' = \boldsymbol{k} + \boldsymbol{e}_1, \\ \lambda_h, & \text{if } k_h < r, \ \boldsymbol{k}' = \boldsymbol{k} + \boldsymbol{e}_2, \\ k_o \mu_o, & \text{if } \boldsymbol{k}' = \boldsymbol{k} - \boldsymbol{e}_1, \\ k_h \mu_h, & \text{if } \boldsymbol{k}' = \boldsymbol{k} - \boldsymbol{e}_2, \\ 0 & \text{in other cases.} \end{cases} \tag{1.40}$$

The steady-state probabilities are determined from the respective SGBE, which is constructed by using relation (1.40). The indicated SGBE has the form like Eqs. (1.31) and (1.37), and its explicit form does not show here. Then, the solution of this SGBE is used to calculate all required QoS metrics of the proposed hybrid access scheme.

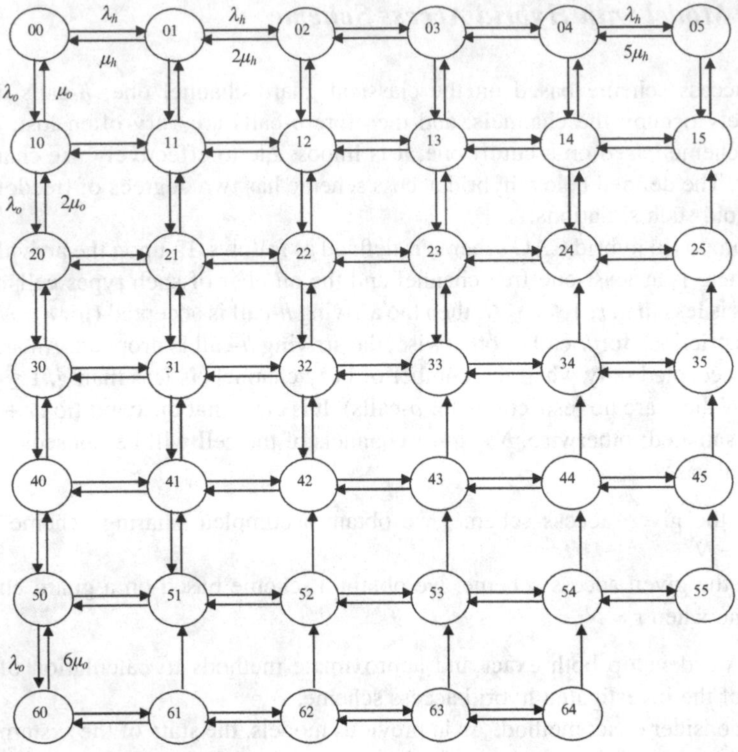

Fig. 1.3 State diagram for the model with hybrid access scheme

The blocking probability of o-calls and channel utilization are calculated like Eqs. (1.34) and (1.35), respectively. However, under this access scheme, the dropping probability of h-calls is defined as follows:

$$P_h = \sum_{\mathbf{k} \in S} p(\mathbf{k})(\delta(k_h, r) + (1 - \delta(k_h, r)\delta(k_o + k_h, N))). \tag{1.41}$$

The method based on the solution of SGBE to calculate QoS metrics is referred to as the exact method. Note that for the real-life networks indicated above, the SGBE have large size, which is why their solution is difficult to calculate. To overcome this obstacle, we developed an approximate method for calculating QoS metrics when using the proposed hybrid access scheme.

The developed method is based on the principles of state space merging of Markov processes [23]. To correctly apply this method, below it is assumed that $\lambda_o << \lambda_h, \mu_o << \mu_h$. Note that these conditions define a regime that commonly occurs in cellular networks with microcells (i.e., when the cell size is small) in which new calls have both longer holding times and significantly lower arrival rates than handover calls. Moreover, as it will be shown below, QoS metrics are determined via loads of heterogeneous traffic (i.e., $v_o = \lambda_o/\mu_o$, $v_h = \lambda_h/\mu_h$) and are independent of specific values of traffic parameters.

By using relationship (1.40), it is easy to show that a given 2-D MC is strongly continuous with respect to the second component and weakly continuous with respect to the first component (for appropriate definitions, see Appendix of the book [29]). Taking into account this fact considers the following splitting of the state space (1.39):

$$S = \cup_{n=0}^{g} S_n, S_n \cap S_m = \varnothing, \quad n \neq m, \tag{1.42}$$

where $S_n = \{\boldsymbol{k} \in S : k_o = n\}$. In other words, we consider the split of the state diagram in Fig. 1.3 along the rows.

Furthermore, state classes S_n are combined into separate merged states $\langle n \rangle$, and the following merging function in state space S is defined:

$$U(\boldsymbol{k}) = \langle n \rangle \text{ if } \boldsymbol{k} \in S_n, \quad n = 0, 1, \ldots, g. \tag{1.43}$$

Function (1.43) determines a merged model that is a 1-D MC with the state space $\Omega = \{\langle n \rangle : n = 0, 1, \ldots, g\}$. Thus, the stationary distribution of the initial model approximately equals (see Appendix of book [29])

$$p(n,m) \approx \rho_n(m)\pi(\langle n \rangle), \ (n,m) \in S_n, \quad n = 0, 1, \ldots, g, \tag{1.44}$$

where $\{\rho_n(m) : (n,m) \in S_n\}$ is the stationary distribution of the split model with state space S_n and $\{\pi(\langle n \rangle) : \langle n \rangle \in \Omega\}$ is the stationary distribution of the merged model.

From the state diagram of the split model with state space S_n (see Fig. 1.3), we conclude that its stationary distribution coincides with an appropriate Erlang's model. Here, we must distinguish two cases: (1) $0 \leq n \leq N - r$ and (2) $N - r + 1 \leq n \leq g$. In case (1), the stationary distribution of all of the split models coincides with that of Erlang's model *M/M/r/0* with load v_n erl (i.e., independent of n), while in case (2) these distributions coincide with those of Erlang's model *M/M/N* $-$ *n/0* with the same load (i.e., dependent on n). Therefore, the stationary distribution of the split model with state space S_n is defined as follows:

Case 1:

$$\rho_n(m) = \frac{v_h^m}{m!} \Big/ \sum_{i=0}^{r} \frac{v_h^i}{i!}, \quad m = 0, 1, \ldots r, \tag{1.45}$$

Case 2:

$$\rho_n(m) = \frac{v_h^m}{m!} \Big/ \sum_{i=0}^{N-r} \frac{v_h^i}{i!}, \quad m = 0, 1, \ldots N - n. \tag{1.46}$$

Then, from Eqs. (1.40), (1.45), and (1.46), the elements of Q-matrix of a merged model $q(\langle n \rangle, \langle n' \rangle), \langle n \rangle, \langle n' \rangle \in \Omega$ are determined as

$$q(\langle n \rangle, \langle n' \rangle) = \begin{cases} \lambda_o, & \text{if } n \leq g - 1 - r,\ n' = n + 1, \\ \lambda_o \alpha(n), & \text{if } n > g - 1 - r,\ n' = n + 1, \\ n\mu_o, & \text{if } n' = n - 1, \\ 0 & \text{in other cases,} \end{cases} \tag{1.47}$$

where $\alpha(n) = \sum_{i=0}^{g-n-1} \rho_n(i)$.

The latter formula allows for the calculation of the stationary distribution of a merged model. It coincides with the appropriate distribution of state probabilities of a 1-D BDP for which transition intensities are determined in accordance with Eq. (1.47). Hence, the desired stationary distribution is calculated as follows:

Case $r \leq g - 1$:

$$\pi(n) = \begin{cases} \frac{v_o^n}{n!} \pi(\langle 0 \rangle), & \text{if } 1 \leq n \leq g - r, \\ \frac{v_o^n}{n!} \prod_{k=g-r}^{n-1} \alpha(k) \pi(\langle 0 \rangle), & \text{if } g - r + 1 \leq n \leq g, \end{cases} \tag{1.48}$$

$$\text{where } \pi(\langle 0 \rangle) = \left(\sum_{n=0}^{g-r} \frac{v_o^n}{n!} + \sum_{n=g-r+1}^{g} \frac{v_o^n}{n!} \prod_{k=g-r}^{n-1} \alpha(k) \right)^{-1}$$

Case $r > g - 1$:

$$\pi(\langle n \rangle) = \frac{v_o^n}{n!} \prod_{k=0}^{n-1} \alpha(k) \pi(\langle 0 \rangle), \tag{1.49}$$

$$\text{where } \pi(\langle 0 \rangle) = \left(\sum_{n=0}^{g} \frac{v_o^n}{n!} \prod_{k=0}^{n-1} \alpha(k) \right)^{-1}.$$

Hereinafter, we assume that $\sum_{i=n}^{m} x_i = 0$ and $\prod_{i=n}^{m} x_i = 1$ if $m < n$.

By using Eqs. (1.45)–(1.49), after some algebraic manipulation, we can obtain the following approximate formulas for calculating the QoS metrics of the model studied:

Case $r \le g-1$:

$$P_o \approx \pi(\langle g\rangle) + \sum_{n=g-r}^{g-1} (\langle n\rangle) \sum_{i=g-n}^{r} \rho_n(i), \tag{1.50}$$

$$C_{\text{u}} \approx \sum_{n=1}^{r} n \sum_{i=0}^{n} \rho_i(n-i)\pi(\langle i\rangle) + \sum_{n=r+1}^{g} n \sum_{i=n-r}^{n} \rho_i(n-i)\pi(\langle i\rangle)$$
$$+ \sum_{n=g+1}^{N} n \sum_{i=n-r}^{g} \rho_i(n-i)\pi(\langle i\rangle), \tag{1.51}$$

Case $r > g-1$:

$$P_o \approx \pi(\langle g\rangle) + \sum_{n=0}^{g-1} \pi(\langle n\rangle) \sum_{i=g-n}^{r} \rho_n(i), \tag{1.52}$$

$$C_{\text{u}} \approx \sum_{n=1}^{g} n \sum_{i=0}^{n} \rho_i(n-i)\pi(\langle i\rangle) + \sum_{n=g+1}^{r} n \sum_{i=0}^{g} \rho_i(n-i)\pi(\langle i\rangle)$$
$$+ \sum_{n=r+1}^{N} n \sum_{i=n-r}^{g} \rho_i(n-i)\pi(\langle i\rangle). \tag{1.53}$$

In both cases $r \le g-1$ and $r > g-1$, the QoS metric P_h is calculated as follows:

$$P_h \approx \rho_0(r) \sum_{n=0}^{N=r} \pi(\langle n\rangle) + \sum_{n=N-r+1}^{g-1} \rho_n(N-n)\pi(\langle n\rangle). \tag{1.54}$$

From formulas (1.50)–(1.54), we conclude that the QoS metrics of the proposed hybrid CAC are determined via loads v_o and v_h of heterogeneous traffic only and are independent of specific values of traffic parameters.

In the special case $r = N$, we obtain CAC based on a classical guard channel scheme. In this case, formulas (1.50)–(1.54) completely coincide with the results obtained in [26], where a great number of computational experiments demonstrate the high accuracy of the proposed approach. In special case $g = r = N$ from Eqs. (1.50)–(1.54), we obtain formulas for CAC based on a complete sharing scheme (see [26] also).

At the end of this subsection, note that adaptation of the proposed hybrid access scheme for use in multiservice cellular networks is straightforward.

1.2.4 Selection of Optimal Values for Parameters of Hybrid Access Scheme

The problem of how to provide a desired QoS level for different call types is scientifically and practically very interesting. The solution of such kind of problems requires some regulated parameters. Thus, in some networks, for which the distribution of channels between cells is fixed, only call admission control parameters can be regulated because controlling loads present a difficult task and one that is sometimes practically impossible.

Thus, we address several problems arising from trying to find optimal parameter values for the proposed in Sect. 1.2.3 hybrid scheme to meet the required QoS level. This scheme has two controlled parameters g and r. To be brief, we address here only two problems of finding the optimal values for the parameters of the proposed hybrid CAC scheme to meet a required QoS level.

Problem 1 It is required to minimize the loss probability of h-calls subject to given restriction to loss probability of o-calls, i.e.,

$$P_h(g,r) \to \min, \\ \text{s.t.}\, P_o(g,r) \leq \varepsilon_o, \tag{1.55}$$

where $\varepsilon_o > 0$ is given value.

Hereinafter, to underline the dependence of functions P_o and P_h on parameters r and g, we specify these parameters in brackets.

Note that solution of the problem (1.55) does not represent any difficulties for moderate size of state space (1.39), i.e., for such models this problem might be solved simply by considering all possible values of the parameters g and r. However, below we propose an effective algorithm to solving similar problems.

First of all, note that to solve this problem and problem 2 (see below), the following unimprovable limits for the investigated functions are useful:

When parameter r is fixed,

$$P_o(N-1,r) \leq P_o(g,r) \leq P_o(N-r,r), \tag{1.56}$$

$$P_h(N-r,r) \leq P_h(g,r) \leq P_h(N-1,r). \tag{1.57}$$

When parameter g is fixed,

$$P_o(g,N-g) \leq P_o(g,r) \leq P_o(g,N), \tag{1.58}$$

$$P_h(g,N) \leq P_h(g,r) \leq P_h(g,N-g). \tag{1.59}$$

Taking into account the monotonic property of functions $P_o(g,r)$ and $P_h(g,r)$ with respect to both parameters and relations (1.56)–(1.59), the following algorithm can be suggested as a solution of the given problem.

Fix the value either g or r. To be concrete, fix the value of g over interval $[1,N]$.

The solution of the problem (1.55) for given value of g, if it exists, is denoted by $(g,r^*(g))$.

Algorithm 1

Step 1. If $\varepsilon_o < P_o(g,N-g)$, then the given problem has no solution.
Step 2. If $\varepsilon_o \geq P_o(g,N)$, then solution of the given problem is $(g,r^*(g))=(g,N)$.
Step 3. If $\varepsilon_o > P_o(g,N)$ and $P_o(g,N-g)\leq \varepsilon_o < P_o(g,N)$, then solution of the given problem is $(g,r^*(g))=(g,r_1)$ where r_1 is determined as solution of the following problem:

$$r_1 = \arg\ \max\{P_o(g,r) \leq \varepsilon_o\}.$$

Then the solution of the investigated problem denoted by (g^*,r^*) is found as $g^*=\arg\min P_h(g,r^*(g))$.

Results of the problem (1.55) for initial data $N=40$ channels $v_o=5$ Erl and $v_h=10$ Erl are shown in Table 1.1.

Problem 2 Let the QoS for different types of calls be measured by the limit values of the blocking and dropping probabilities, i.e., their upper bounds are set:

$$P_o(g,r) \leq \varepsilon_o, \tag{1.60}$$

$$P_h(g,r) \leq \varepsilon_h, \tag{1.61}$$

where $\varepsilon_o > 0$ and $\varepsilon_h > 0$ are given values.

The optimization problem is formulated as follows: it is required to find extreme values of parameters g and r such that conditions (1.60) and (1.61) are met.

Again taking into account the monotonic property of functions $P_o(g,r)$ and $P_h(g,r)$ with respect to both parameters and relations (1.56)–(1.59), the following algorithm can be suggested to the solution of the given problem.

Fix the value of g, $g\in[1,N]$.

Algorithm 2

Step 1. If $\varepsilon_o < P_o(g,N-g)$ or $\varepsilon_h < P_h(g,N)$, then the problem has no solution.
Step 2. If $\varepsilon_o > P_o(g,N)$ and $\varepsilon_h > P_h(g,N-g)$, then the solutions are the pairs (g,r), where $r\in[N-g,N]$.
Step 3. If $\varepsilon_o > P_o(g,N)$ and $P_h(g,N)\leq \varepsilon_h \leq P_h(g,N-g)$, then the following problem is to be solved:

Table 1.1 Solution results for the problem (1.55)

ε_o	E-2	E-3	E-4	E-5	E-6	E-7	E-8	E-9	E-10	E-11
(g^*, r^*)	(26,40)	(29,40)	(32,40)	(35,40)	(38,40)	(40,40)	(40,24)	(40,20)	(40,17)	(40,15)
min P_h	2.45E-8	3.3E-8	3.97E-8	4.1E-8	4.1E-8	4.1E-8	7.32E-5	0.002	0.013	0.036

$$r_h = \arg\min\{P_h(g,r) \leq \varepsilon_h\}.$$

The solutions are the pairs (g, r), where $r \in [r_h, N]$.

Step 4. If $\varepsilon_h > P_h(g, N-g)$ and $P_o(g, N-g) \leq \varepsilon_o \leq P_o(g, N)$, then the following problem is to be solved:

$$r_o = \arg\max\{P_o(g,r) \leq \varepsilon_o\}.$$

The solutions are the pairs (g, r), where $r \in [N-g, r_o]$.

Step 5. If $P_o(g, N-g) \leq \varepsilon_o \leq P_o(g, N)$ and $P_h(g, N) \leq \varepsilon_h \leq P_h(g, N-g)$, then the problem has no solution if $r_o < r_h$; otherwise, the solutions are the pairs (g, r), where $r \in [r_h, r_o]$.

Note that particularly for the solutions of the problems in steps 3 and 4, a dichotomy method can be used due to the monotonic property of the investigated functions.

Combining the solutions of the problem for different possible values of parameter g, we find a range of parameters g and r for which both conditions (1.60) and (1.61) are met.

Some solutions of problems (1.60) and (1.61) for the same initial data, i.e., for $N = 40$ channels, $v_o = 5$ Erl and $v_h = 10$ Erl, $\varepsilon_o = 10^{-3}$, are presented in Figs. 1.4 and

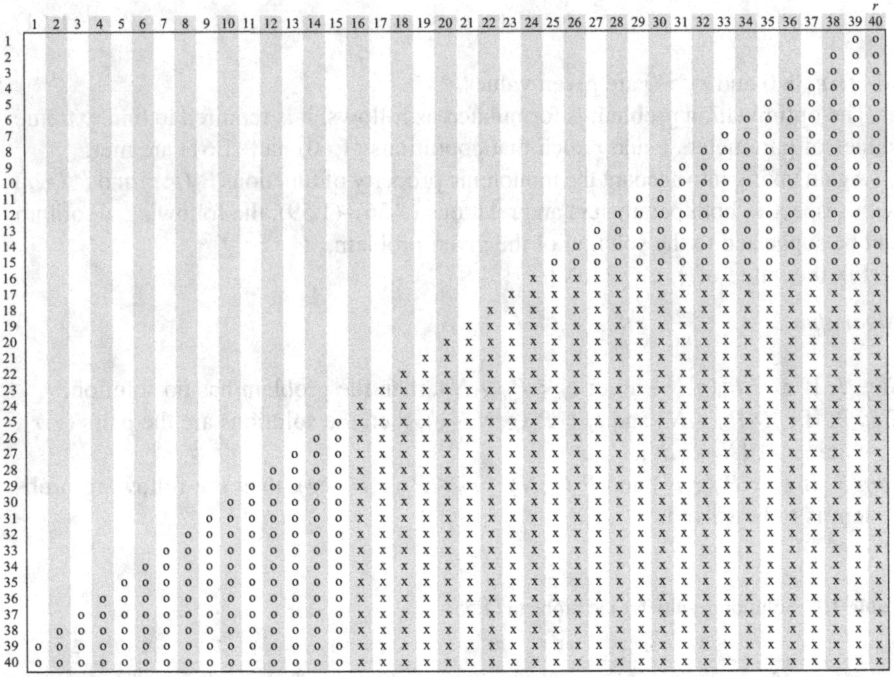

Fig. 1.4 Solution set for the problems (1.60), (1.61) in case $\varepsilon_h = 10^{-4}$

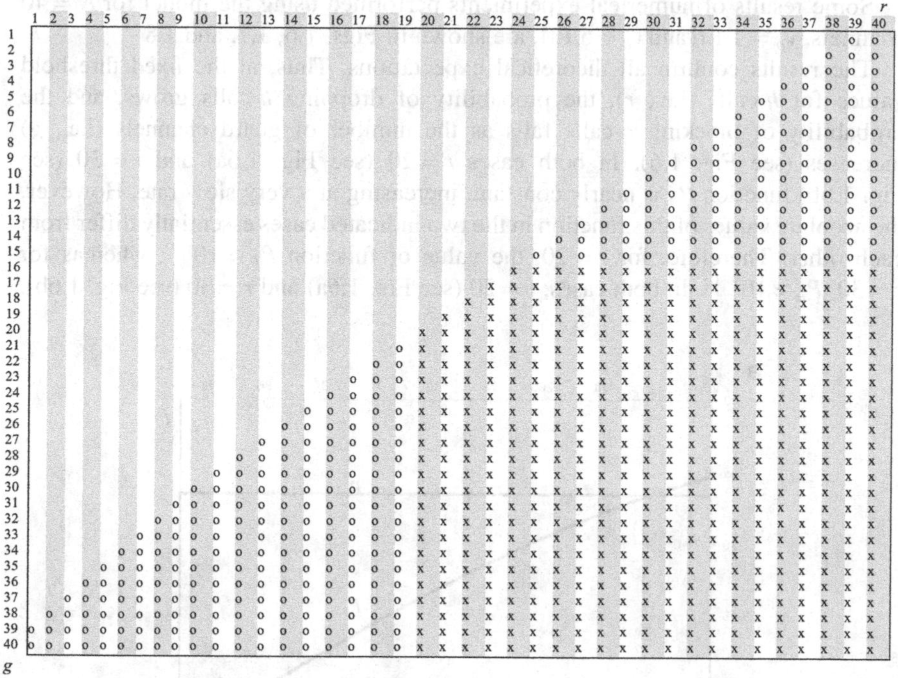

Fig. 1.5 Solution set for the problems (1.60), (1.61) in case $\varepsilon_h = 10^{-6}$

1.5, where symbol 0 indicates no solution and symbol x indicates that this point is a solution of the given problem.

From these figures, we conclude that a stronger restriction of either one or both functions P_υ and P_h leads to a reduction in the solution set of the indicated problem.

1.3 Numerical Results

In the literature, there are large number numerical experiments which were carried out by means of algorithms from Sect. 1.1. Therefore, here we will not consider the results related to classical models of multidimensional models. Here only we will notice that the presence of explicit formulas to calculate the QoS metrics essentially facilitates the solution of the problem.

Below, we consider results of numerical experiments which are carried out by using algorithms developed in Sect. 1.2. The mentioned algorithms suggested allow one to study the behavior of the QoS metrics of the proposed hybrid scheme over all admissible ranges of their structural and load parameters.

For brevity, only results regarding the dependence of QoS metrics on the parameters of the hybrid CAC scheme (i.e., g and r) are presented in detail.

Some results of numerical experiments performed using the model for $N=40$ channels, $v_o=1$ Erl and $v_h=5$ Erl, are shown in Figs. 1.6, 1.7, and 1.8.

The results confirm all theoretical expectations. Thus, at the fixed threshold values for *h*-calls (i.e., *r*), the probability of dropping *h*-calls grows, and the probability of blocking *o*-calls falls as the number of guard channels (i.e., *g*) increases (see Fig. 1.6). In both cases $r=20$ (see Fig. 1.6a) and $r=30$ (see Fig. 1.6b) function P_h is nearly constant, increasing at a very slow rate. However, the absolute values of this function in the two indicated cases essentially differ from each other. Therefore, for $r=20$, the value of function $P_h \approx 10^{-6}$, whereas for $r=30$, $P_h \approx 10^{-13}$. In both cases, $r=20$ (see Fig. 1.6a) and $r=30$ (see Fig. 1.6b),

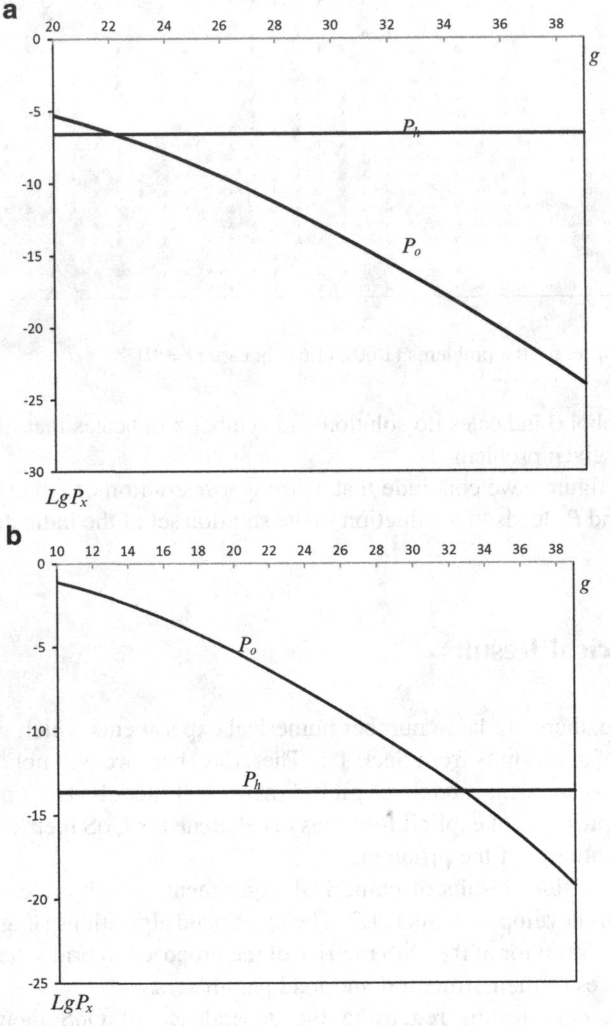

Fig. 1.6 Loss probabilities versus $g, a-r=20$; $b-r=30$

Fig. 1.7 Loss probabilities versus $r, a - g = 20$; $b - g = 30$

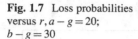

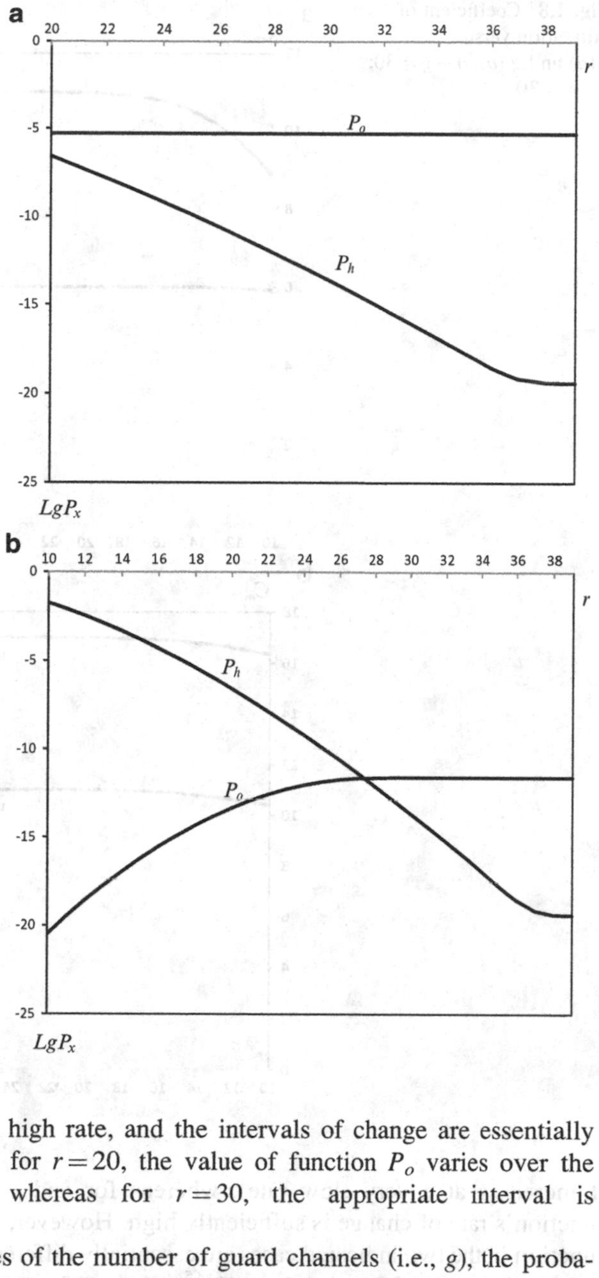

function P_o decreases at a high rate, and the intervals of change are essentially different from each other: for $r = 20$, the value of function P_o varies over the interval $[10^{-25}, 10^{-5}]$, whereas for $r = 30$, the appropriate interval is $[10^{-20}, 10^{-2}]$.

Similarly, at fixed values of the number of guard channels (i.e., g), the probability of dropping h-calls falls (see Fig. 1.7a), and the probability of blocking o-calls grows (see Fig. 1.7b) as the value of threshold for h-calls (i.e., r) increases. Unlike in the previous graphs, for $g = 20$ (see Fig. 1.7a) and $g = 30$ (see Fig. 1.7b), function P_h falls at high rates. For $g = 20$ (see Fig. 1.7a), function P_o is nearly constant (i.e.,

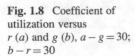

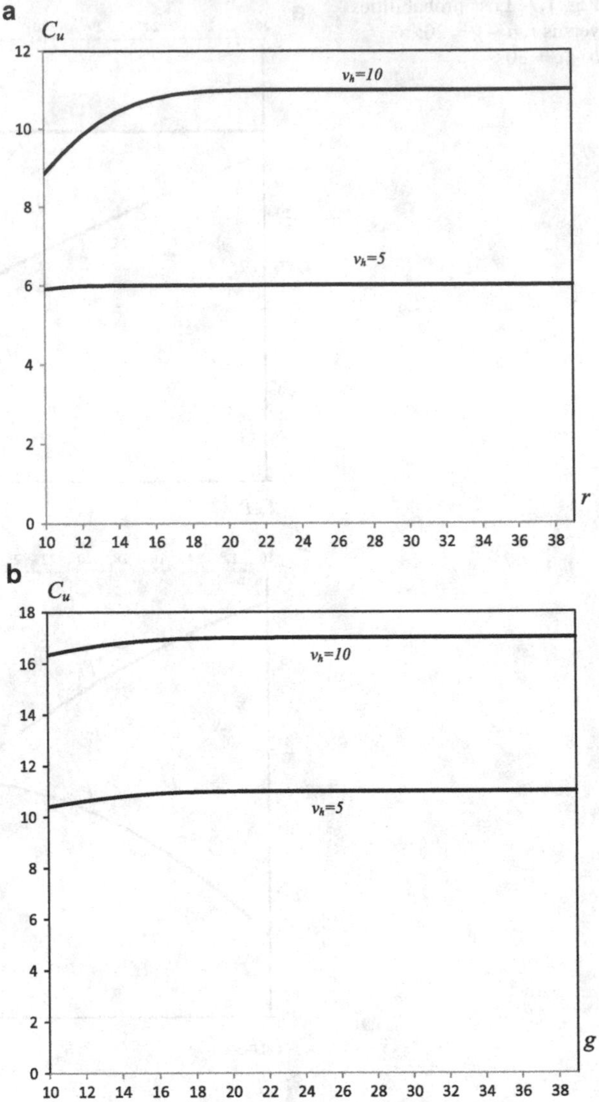

Fig. 1.8 Coefficient of utilization versus r (*a*) and g (*b*), $a - g = 30$; $b - r = 30$

it increases at a very slow rate), whereas for and $g = 30$ (see Fig. 1.7b), the function's rate of change is sufficiently high. However, the absolute values of this function in the two indicated cases are essentially different from each other, i.e., for $g = 20$, the value of function $P_o \approx 10^{-5}$, whereas for $g = 30$, $P_o \in [10^{-20}, 10^{-12}]$.

The channel utilization metric is an increasing function with respect to both parameters g and r (see Fig. 1.8a, b). However, for given traffic loads, this metric has no high rate of change with respect to parameters g and r. As was expected, function C_u is an increasing function with respect to the loads of heterogeneous calls, which holds for all QoS metrics.

Another objective of the numerical experiments was to estimate the accuracy of the proposed formulas. For small models (i.e., when the number of channels is not too large), the exact values of the QoS metrics are determined by the solving appropriate system of global balance equations. Note that the approximate values of QoS metrics are nearly identical to their exact values when the accepted assumption about the ratios of the load parameters of o- and h-calls is valid. Moreover, the approximate formulas are highly accurate even when the accepted assumption is not valid, which might be explained by the fact that, as mentioned above, QoS metrics are determined via v_o and v_h and are independent of specific values of traffic parameters. For brevity, results that compare the exact and approximate values of QoS metrics are considered for special case, i.e., for model with CAC based on classical guard channel scheme [26].

For the indicated special case (i. e., $r = N$) from Eqs. (1.50)–(1.54), we have

$$P_h \approx \sum_{i=0}^{g} E_B(v_h, N - i)\pi(\langle i \rangle), \tag{1.62}$$

$$P_o \approx \sum_{i=0}^{g} \sum_{j=g-i}^{N-i} \rho_i(j)\pi\langle 0 \rangle. \tag{1.63}$$

In last formulas the following notations are used:

$E_B(v, N)$—Erlang's B-formula, i.e., $E_B(v, N) = (v^N/N!)(\sum_{i=0}^{N}(v^i/I!))^{-1}$;

$\rho_i(j) = \frac{v_h^j}{j!}\rho_i(0)$, $i = 0, 1, \ldots, g; j = 0, 1, \ldots, N - i$, where $\rho_i(0) = \left(\sum_{j=0}^{g} \frac{v_h^j}{j!}\right)^{-1}$;

$$\pi(\langle i \rangle) = \frac{v_o^i}{i!} \prod_{j=1}^{i} \Lambda(j)\pi(\langle 0 \rangle), \quad i = 1, 2, \ldots, g,$$

where

$$\pi(\langle 0 \rangle) = \left(\sum_{i=0}^{g} \frac{v_o^i}{i!} \prod_{j=1}^{i} \Lambda(j) \right)^{-1},$$

$$\Lambda(i + 1) = \rho_i(0) \sum_{j=0}^{g-i-1} \frac{v_h^j}{j!}, \quad i = 0, 1, \ldots, g - 1.$$

The results achieved with this algorithm nearly coincide with those from exact calculations under the indicated above conditions: $\lambda_o << \lambda_h$, $\mu_o << \mu_h$. Under these conditions, the exact values of the QoS metrics for a model of moderate dimensions are calculated with SGEE.

In [11] also introduced an approximation algorithm for solving this problem, which was based on the following heuristic considerations. The different channel occupation times are replaced with their averages, and the traffic intensities are replaced by their loads; that is, the following approximations are made: $\mu_o = \mu_h = 1$, $\lambda_o = v_o$, $\lambda_h = v_h$. The paper [11] proposed the following approximate formulas for calculating the QoS metrics:

$$P_h \approx \rho_N, \tag{1.64}$$

$$P_o \approx \sum_{i=g}^{N} \rho_i, \tag{1.65}$$

where

$$\rho_i = \begin{cases} \dfrac{v^i}{i!}\rho_0, & \text{if } i \le g, \\ \dfrac{v^g v_h^{i-g}}{i!}\rho_0, & \text{if } g+1 \le i \le N, \end{cases} \tag{1.66}$$

$$\rho_0 = \left(\sum_{i=0}^{g} \frac{v^i}{i!} \sum_{i=g+1}^{N} \frac{v^g v_h^{i-g}}{i!} \right)^{-1}, \quad v = v_0 + v_h. \tag{1.67}$$

The authors [11] noted that it is impossible to measure the accuracy of the proposed formulas analytically; therefore, they demonstrated the high accuracy of these formulas using simulations.

Another way of transforming an adequate two-dimensional model to an approximate one-dimensional model is as follows. Different average channel occupation times are replaced with weighted average $\mu = v/\lambda$, where $\lambda = \lambda_o + \lambda_h$. Such an approach to the problem solution is called traditional [11]. In this case, for QoS parameter calculations, they also use formulas (1.64) and (1.65); however, in formulas (1.66) and (1.67), parameters v_h and v_o are defined as follows: $v_h = \lambda_h/\mu$, $v_o = \lambda_o/\mu$.

Paper [11] also showed low accuracy of this traditional approach, especially when μ_o and μ_h vary greatly. As was expected, results of last two approaches coincide at $\mu_o = \mu_h$.

Tables 1.2, 1.3, and 1.4 contain results of numerical experiments for the above three methods. The comparison shows that the results of papers [11] and [26] do not differ significantly, even when condition $\lambda_o << \lambda_h$, $\mu_o << \mu_h$ does not hold true (see Tables 1.3 and 1.4).

Table 1.2 Comparison of different algorithms for the case $N=10$, $\lambda_o=1$, $\lambda_h=5$, $\mu_o=1$, $\mu_h=5$

	P_o			P_h		
m	[26]	[11]	Traditional	[26]	[11]	Traditional
1	0.7310586	0.7746003	0.8374876	3.4676E-07	1.24228E-07	8.88776E-06
2	0.4621171	0.4891989	0.5577866	1.1128E-06	1.87684E-07	9.67381E-06
3	0.2516532	0.2588479	0.2998047	3.1089E-06	3.26787E-07	1.10285E-05
4	0.1163676	0.1153544	0.1327018	7.2415E-06	6.15877E-07	1.29415E-05
5	0.0453591	0.0435004	0.0492806	1.3815E-05	1.20496E-06	1.54023E-05
6	0.0149759	0.0140255	0.0156344	2.1738E-05	2.39303E-06	1.84346E-05
7	0.0042446	0.0039216	0.0043088	2.8994E-05	4.77667E-06	2.21056E-05
8	0.0010474	0.0009644	0.0010449	3.4104E-05	9.54873E-06	2.65221E-05
9	0.0002249	0.0002101	0.0002227	3.6928E-05	1.90954E-05	3.18253E-05

Table 1.3 Comparison of different algorithms for the case $N=10$, $\lambda_o=5$, $\lambda_h=1$, $\mu_o=5$, $\mu_h=1$

	P_o			P_h		
m	[26]	[11]	Traditional	[26]	[11]	Traditional
1	0.73105860	0.77460030	0.7035873	3.4676E-07	1.24228E-07	8.29989E-12
2	0.46211714	0.48919887	0.42770452	1.1128E-06	1.87684E-07	3.20498E-11
3	0.25165324	0.25884799	0.22508059	3.1089E-06	3.26787E-07	1.5623E-10
4	0.11636757	0.11535437	0.10127703	7.2415E-06	6.15877E-07	8.58268E-10
5	0.04535909	0.04350037	0.03875433	1.3815E-05	1.20496E-06	4.9833E-09
6	0.01497587	0.01402552	0.01267733	2.1738E-05	2.39303E-06	2.95839E-08
7	0.00424458	0.00392164	0.00358918	2.8994E-05	4.77667E-06	1.76973E-07
8	0.00104741	0.00096442	0.00089234	3.4104E-05	9.54873E-06	1.06105E-06
9	0.00022489	0.00021005	0.00019732	3.6928E-05	1.90954E-05	6.36523E-06

Table 1.4 Comparison of different algorithms for the case $N=10$, $\lambda_o=\lambda_h=5$, $\mu_o=\mu_h=5$

	P_o			P_h		
m	[26]	[11]	Traditional	[26]	[11]	Traditional
1	0.73105860	0.77460030	0.7746003	3.4676E-07	1.24228E-07	1.24228E-07
2	0.46211714	0.48919887	0.48919887	1.1128E-06	1.87684E-07	1.87684E-07
3	0.25165324	0.25884799	0.25884799	3.1089E-06	3.26787E-07	3.26787E-07
4	0.11636757	0.11535437	0.11535437	7.2415E-06	6.15877E-07	6.15877E-07
5	0.04535909	0.04350037	0.04350037	1.3815E-05	1.20496E-06	1.20496E-06
6	0.01497587	0.01402552	0.01402552	2.1738E-05	2.39303E-06	2.39303E-06
7	0.00424458	0.00392164	0.00392164	2.8994E-05	4.77667E-06	4.77667E-06
8	0.00104741	0.00096442	0.00096442	3.4104E-05	9.54873E-06	9.54873E-06
9	0.00022489	0.00021005	0.00021005	3.6928E-05	1.90954E-05	1.90954E-05

1.4 Conclusion

There are fundamental monographs that are not only focused to foundations of queueing theory but also to advanced up-to-date methods to application in telecommunication networks [13, 14, 16, 20, 21, 28]. In these books detailed analysis of main problems in applied queueing theory is considered.

Classical Erlang's model [7] has been widely used for more than hundred years for the analysis of various communication systems. Its scientific value has essentially increased after the Kovalenko's classical result [24]: the stationary distribution of this model has been proved to be invariant with respect to the distribution function of holding time for fixed mean value. This fact has essentially expanded application area of the MRQ models since the exponential distribution function of a holding time causes serious disputes among researchers and specialists in telecommunication area. Insensitivity of the stationary distribution was confirmed later by other authors (see, e.g., [3] and the bibliography therein).

Since modern telecommunication networks are multiservice, models with heterogeneous calls (i.e., multidimensional Erlang's models) are actively investigated. Here we will not discuss the known results which are related to classical multidimensional Erlang's models since these results can be found in the abovementioned books. But instead we briefly consider the review of the applied works in cellular networks.

Almost in all early works devoted to modeling of cellular networks, it was supposed that both handover and new calls are identical in terms of channel occupancy time (see [1, 2, 32] and references therein). However, this assumption is unrealistic [8–11, 19, 25, 26, 35]. So, let us briefly review the results of previous studies of models of cellular networks with nonidentical channel occupancy times for calls of different types.

One of the first works to investigate such models was [11]. This paper examined models featuring various GC schemes for h-calls or establishing thresholds (limits) for o-calls. For the latter type of model, simple formulas are easily developed for calculating the loss probability of h-calls as well as the blocking probability of o-calls. The development of simple formulas is possible because there is an explicit (multiplicative) solution to an appropriate SGBE for the state probabilities. However, there is no closed form solution to the corresponding equations for models with CAC based on the GC scheme; therefore, the authors suggested approximation formulas for QoS calculations based on some heuristic considerations. The accuracy of the suggested formulas was determined by simulation. The authors also suggested another approximation scheme, which they dubbed traditional and in which different occupancy times are replaced by a single (unified) time by using some procedure. It was shown that the traditional scheme gives a rough approximation, especially as the occupancy times of different calls differ substantially. The common idea underlying both approximation methods is the unification of the channel occupancy times of heterogeneous calls. A similar approximation approach based on this idea is proposed in [37].

In [8], the results of the paper [11] were generalized to multimedia networks. In this study, the decision to accept arriving calls was determined by probabilities that depend on the number of busy channels or the number of ongoing calls belonging to a given class. However, in [8], the proposed schemes were analyzed under the assumptions that all traffic required the same number of channels and their channel occupancy times were identical. In [34], these CAC schemes were generalized to models in which different types of calls require an arbitrary number of channels (i.e., multi-rate models) and exhibit service times that are not identical. This last paper also provided a detailed review of known results in this direction.

A generalized reservation scheme called fractional guard channels (FGC) and its own special case, i.e., uniform fractional guard channel (UFGC) was considered in [12, 15, 33] as well. In [18] an analytical approach for calculating models of a wireless network with nonidentical channel occupancy times for *h*- and *o*-calls in which the CAC are based on GC, FGC, and UFGC schemes is developed. In this paper, it is proved that the formulas suggested in [11, 37] are not merely approximations but are exact if the model satisfies a local balance condition; otherwise, these formulas are approximations. Method to finding the number of guard channels which is adaptive to instantaneous cell traffics is proposed in [38]. Note that in [38] different channel holding time for new and handoff calls takes into account, i.e., to model the investigated cell 2-D MC is used; state probabilities of the appropriate 2-D MC are determined by using the Gauss–Seidel iterative algorithm.

A detailed review of mathematical methods used to study the QoS metrics of wireless networks with different CAC schemes may be found in [5].

Problems relating to analysis and optimization of a new hybrid access scheme examined in this chapter are investigated in [27]. Although here we studied mono-service networks to simplify the described models and corresponding calculations, the results achieved can be adapted to multiservice networks.

References

1. Ahmed M, Yanikomeroglu H (2005) Call admission control in wireless networks. IEEE Commun Surv 7(1):50–69
2. Beigy H, Meybody MR (2003) Used-based call admission control policies for cellular mobile systems: a survey. J CSI Comput Sci Eng 10:45–58
3. Burman DY, Lehoczky JP, Lim Y (1984) Insensitivity of blocking probabilities in a circuit switched network. J Appl Probab 21(4):850–859
4. Buzen JP (1973) Computational algorithms for closed networks with exponential servers. Commun Assoc Comput Mach 16(9):527–531
5. Cruz-Perez FA, Toledo-Marin R, Hernandez-Valdez G (2011) Approximated mathematical methods of guard-channel-based call admission control in cellular networks. In: Melikov A (ed) Cellular networks: positioning, performance analysis, reliability. Intech, India, pp 151–168
6. Delaire M, Hebuterne G (1997) Call blocking in multi-services systems on one transmission link. In: 5th international workshop on performance model and evaluation of ATM networks, July, UK, pp 253–270

7. Erlang AK (1909) The theory of probabilities and telephone conversations. Nyt Tidsskrift for Matematik B-20:33–39
8. Fang Y (2003) Thinning schemes for call admission control in wireless networks. IEEE Trans Comput 52(5):371–382
9. Fang Y, Chlamtac I (1999) Teletraffic analysis and mobility modelling for PCS networks. IEEE Trans Commun 47:1062–1072
10. Fang Y, Chlamtac I, Lin YB (1998) Channel occupancy times and handoff rate for mobile computing and PCS networks. IEEE Trans Comput 47:679–692
11. Fang Y, Zhang Y (2002) Call admission control schemes and performance analysis in wireless mobile networks. IEEE Trans Veh Technol 51(2):371–382
12. Firouzi Z, Beigy H (2009) A new call admission control scheme based on new call bounding and thinning II schemes in cellular mobile networks. In: Proceedings of IEEE conference on electro/information technology, June 2009, pp 40–45. doi:978-1-4244-3355-1/09
13. Gelenbe E, Pujolle G (1998) Introduction to queueing networks. Wiley, New York
14. Gnedenko BV, Kovalenko IN (1989) Introduction to queueing theory. Birkhauser, Boston
15. Goswami V, Swain PK (2012) Analysis of finite population limited fractional guard channel call admission control in cellular networks. Procedia Eng 30:759–766
16. Gross D, Harris C (1985) Fundamentals of queueing theory. Wiley, New York
17. Kaufman JS (1981) Blocking in shared resource environment. IEEE Trans Commun 10 (2):1474–1481
18. Kim HM, Melikov A, Fattakhova M, Kim CS (2012) An analytical approach to the analysis of guard-channel-based call admission control in wireless cellular networks. J Appl Math 12: Article ID 676582, 14 pages, doi:10.1155/2012/676582
19. Kim K (2008) A computation method in performance evaluation in cellular communication network under general distribution model. J KSIAM 12(2):119–131
20. Kleinrock L (1975) Queueing systems, Vol. 1. Theory. Wiley, New York
21. Kleinrock L (1976) Queueing systems, Vol. 2. Computer applications. Wiley, New York
22. Kolmogorov A (1936) Zum theorie der Markoffschen ketten. Mathematische Annalen B112:155–160
23. Korolyuk VS, Korolyuk VV (1999) Stochastic model of systems. Kluwer, Boston
24. Kovalenko IN (1963) On independence of stationary distributions of the form of the service time distribution law. Probl Pered Inform 11:104–113
25. Leoung CW, Zhuang W (2003) Call admission control for wireless personal communications. Comput Commun 26(6):522–541
26. Melikov AZ, Babayev AT (2006) Refined approximations for performance analysis and optimization of queuing model with guard channels for handovers in cellular networks. Comput Commun 29(9):1386–1392
27. Melikov AZ, Ponomarenko LA (2014) Methods for analysis and optimization of a new access strategy in cellular communication networks. J Autom Inform Sci 46(3):70–82
28. Neuts M (1984) Matrix-geometric solutions in stochastic models: an algorithmic approach. John Hopkins University Press, Baltimore
29. Ponomarenko L, Kim CS, Melikov A (2010) Performance analysis and optimization of multi-traffic on communication networks. Springer, Heidelberg
30. Pioro M, Lubacs J, Korner U (1990) Traffic engineering problems in multi-service circuit-switched networks. Comput Netw ISDN Syst 1–5:127–136
31. Roberts JW (1981) A service system with heterogeneous user requirements application to multi-service telecommunication systems. In: Pujolle G (ed) Performance of data communication systems and their application. North Holland, Amsterdam, pp 423–431
32. Schneps-Schneppe M, Iversen VB (2012) Call admission control in cellular networks. In: Ortiz JH (ed) Mobile networks. Intech, India, pp 111–136
33. Vazquez-Avilla JL, Cruz-Perez FA, Ortigoza-Guerrero L (2006) Performance analysis of fractional guard channel policies in mobile cellular networks. IEEE Trans Wireless Commun 5(2):301–305

34. Wang X, Fan P, Pan Y (2008) A more realistic thinning scheme for call admission control in multimedia wireless networks. IEEE Trans Comput 57(8):1143–1148
35. Wei L, Fang Y (2008) Performance evaluation of wireless cellular networks with mixed channel holding times. IEEE Trans Wireless Commun 7(6):2154–2160
36. Wolff RW (1992) Poisson arrivals see time averages. Oper Res 30(2):223–231
37. Yavuz EA, Leung VCM (2006) Computationally efficient method to evaluate the performance of guard-channel-based call admission control in cellular networks. IEEE Trans Veh Technol 55(4):1412–1424
38. Zeng H, Chlamtac I (2003) Adaptive guard channel allocation and blocking probability estimation in PCS networks. Comput Netw 43:163–176

Chapter 2
Analytical Methods for Analysis of Integral Cellular Networks

The classical theory of multidimensional queueing systems is based on many assumptions, one of them is basic: one call—one channel. However, in modern integral (multiservice) communication networks, this assumption is not carried out. Thus, in them, for example, a video information requires the wider band in a digital transmission line than data or voice information. In the teletraffic theory the calls, requiring a large number of channels in a transmission line, are called wideband, and the calls that require a smaller number of channels—narrowband. As it was noted in the previous chapter, the multi-flow system, in which heterogeneous calls require for simultaneous maintenance a random number of channels, is called multi-rate queue (MRQ).

Since in MRQ with inelastic calls in the absence of the required number of free channels call service cannot be started, one would expect that wideband calls will be lost more often than narrowband. Therefore, in such systems to maintain the quality of service (QoS) of heterogeneous calls at the desired level, it is necessary to determine the appropriate CAC scheme.

The alternative way for satisfying the desired QoS level is determining the appropriate schemes to partition of common pool of channels among heterogeneous calls.

In this chapter the new method to study the MRQ model with the randomized access scheme is proposed, and on its basis the efficient algorithms to calculate the QoS metrics in specific telecommunication network models are developed. In addition both isolated and virtual schemes to partition of common pool of channels in integral telecommunication networks are proposed, and exact formulas to calculate the QoS metrics of such networks are obtained. The results of numerical experiments, performed with the help of the developed algorithms, and the meaningful analysis of these results are given.

A. Melikov, L. Ponomarenko, *Multidimensional Queueing Models in Telecommunication Networks*, DOI 10.1007/978-3-319-08669-9_2

2.1 Model of Multi-rate Queue with Randomized Access Scheme

Let the input of the multichannel system, containing $N > 1$ channels, receive a Poisson stream of heterogeneous calls with the total intensity Λ. Every new incoming call with the probability σ_i requires for service simultaneously b_i channels, $1 \leq b_i \leq N$, $i = 1, \ldots, K$; meanwhile $\sigma_1 + \ldots + \sigma_k = 1$. It is believed that at the moment of call arrival the number of channels, requested by it for service, is known. Then, subject to the known properties of the Poisson flow, one can argue that at the input of N channel system K types of the Poisson streams of calls enter, the intensity of the ith stream being $\lambda_i = \Lambda \sigma_i$, $i = 1, \ldots, K$; meanwhile the calls of the ith type (i-calls) require simultaneously b_i channels, $i = 1, \ldots, K$. The holding time of the i-calls is a random variable subjected to an exponentially distribution law with the parameter μ_i, $i = 1, \ldots, K$.

The system uses a randomized scheme for access. For this, the access matrix of the dimension $K \times N$ is determined, in which elements define the rules of reception of heterogeneous calls, depending on their type and the number of busy channels. More precisely, the element $\alpha_i(n)$ of the matrix indicates the probability of reception of the i-call for service, if at the time of its arrival the number of busy channels equals n; with the complementary probability $1 - \alpha_i(n)$, this call is lost. In this model interruption of service process of the call of any type is not allowed, i.e., it is assumed that $\alpha_i(n) = 0$ for any $i = 1, \ldots, K$, if $n > N - b_i$. Note that the condition $b_i = b_j$ at $i \neq j$ not in the least means that the equality $\alpha_i(n) = \alpha_j(n)$ is to be fulfilled.

Let us consider the problem of determining the QoS metrics of the studied model while using the proposed access scheme. The main QoS metrics are stationary loss (blocking) probability of i-calls (PB_i, $i = 1, \ldots, K$) and the average number of the busy channels (N_{av}).

The state of the system at arbitrary time instant is described by the K-dimensional vector $\boldsymbol{m} = (m_1, \ldots, m_K)$, where m_i is a number of the i-calls in the system (i.e., in the channels), $i = 1, \ldots, K$. In other words, the functioning of the given MRQ is described by K-dimensional Markov chain with the following state space:

$$S = \{\boldsymbol{m} : m_i = 0, 1, \ldots, [N/b_i]; (\boldsymbol{m}, \boldsymbol{b}) \leq N\}, \tag{2.1}$$

where $\boldsymbol{b} = (b_1, \ldots, b_k)$; $[x]$ is an integer part of x; $(\boldsymbol{m}, \boldsymbol{b})$ is a scalar product of the vectors $\boldsymbol{m}$ and $\boldsymbol{b}$.

Note that from this scheme one can obtain in particular cases the well-known deterministic access schemes. Let us consider some of them:

1. If $\alpha_i(n) = 1$ for every $i = 1, \ldots, K$ at $n \leq N - b_i$, then one gets the model with full available group of channels, i.e., model with the complete sharing (CS) scheme [13, 29].
2. Assume that the parameters $\alpha_i(n)$ for every $i = 1, \ldots, K$ are defined as follows:

$$\alpha_i(n) = \begin{cases} 1 & \text{if } n \le N - b, \\ 0 & \text{in other cases,} \end{cases} \tag{2.2}$$

where $b = \max\{b_i : i = 1, \ldots, K\}$. Then one gets the complete sharing with equalization (CSE) scheme [7], i.e., the received call of any type is served, if at this moment the number of free channels is not less than b.

3. Assume that the parameters $\alpha_i(n)$ for every $i = 1, \ldots, K$ are defined as follows:

$$\alpha_i(n) = \begin{cases} 1 & \text{if } n \le N - b_i - r_i, \\ 0 & \text{in other cases,} \end{cases} \tag{2.3}$$

where $0 \le r_i \le N - b_i$. Then one gets the trunk reservation (TR) scheme [28], i.e., if at the time of receiving the i-call the number of free channels is not less than $b_i + r_i$, then it is served; otherwise the received call is lost with probability 1. The parameter r_i is called the backup parameter of the channels for i-calls, $i = 1, \ldots, K$.

Let us state the proposed method of solving the problem. Transitions between the states $\boldsymbol{m}$ and $\boldsymbol{m}' \in S$ occur only at the moment of receiving calls and their leaving the system after completion of service. In view of this, the nonnegative elements of the Q-matrix of the given multidimensional Markov chain are determined from the following relationships:

$$q(\boldsymbol{m}, \boldsymbol{m}') = \begin{cases} \lambda_i \alpha_i((\boldsymbol{m}, \boldsymbol{b})) & \text{if } \boldsymbol{m}' = \boldsymbol{m} + \boldsymbol{e}_i, \\ m_i \mu_i & \text{if } \boldsymbol{m}' = \boldsymbol{m} - \boldsymbol{e}_i, \\ 0 & \text{in other cases,} \end{cases} \tag{2.4}$$

where $\boldsymbol{m}, \boldsymbol{m}' \in S$, $\boldsymbol{e}_i$ is the ith orthogonal vector in K-dimensional Euclidean space, $i = 1, \ldots, K$.

For any positive values of the parameters of incoming traffics, all the states are communicating and, consequently, the system is ergodic. Let us denote the stationary probability of state $\boldsymbol{m} \in S$ as $p(\boldsymbol{m})$. The desired QoS metrics are determined in terms of the steady-state probabilities. Thus the stationary probability of blocking the i-calls is calculated as follows:

$$\text{PB}_i = \sum_{n=0}^{N} (1 - \alpha_i(n)) \sum_{\boldsymbol{m} \in S_n} p(\boldsymbol{m}), \quad i = 1, \ldots, K, \tag{2.5}$$

where $S_n = \{\boldsymbol{m} \in S : (\boldsymbol{m}, \boldsymbol{b}) = n\}$, $n = 0, 1, \ldots, N$, i.e., the sets S_n combine the microstates from state space (2.1) with the same number of busy channels.

Note 2.1 From formula (2.5), one gets that if $\alpha_i(n) = \alpha_j(n)$ at $b_i = b_j$ for some i, j, $i \ne j$, then $\text{PB}_i = \text{PB}_j$, for any values of the model load parameters.

The average number of busy channels is defined as

$$N_{\text{av}} = \sum_{n=1}^{N} n \sum_{\boldsymbol{m} \in S_n} p(\boldsymbol{m}). \tag{2.6}$$

The main problem in finding the QoS metrics (2.5) and (2.6) is the calculation of $p(\boldsymbol{m})$, $\boldsymbol{m} \in S$, which satisfies the corresponding system of global balance equations (SGBE):

$$\left(\sum_{i=1}^{K} \lambda_i \alpha_i((\boldsymbol{m}, \boldsymbol{b})) I((\boldsymbol{m}, \boldsymbol{b}) \leq N - b_i) + \sum_{i=1}^{K} m_i \mu_i \right) p(\boldsymbol{m})$$
$$= \sum_{i=1}^{K} \lambda_i \alpha_i((\boldsymbol{m} - \boldsymbol{e}_i, \boldsymbol{b})) p(\boldsymbol{m} - \boldsymbol{e}_i) I(m_i > 0)$$
$$+ \sum_{i=1}^{K} (m_i + 1) p(\boldsymbol{m} + \boldsymbol{e}_i) I(\boldsymbol{m} + \boldsymbol{e}_i \in S), \tag{2.7}$$

$$\sum_{\boldsymbol{m} \in S} p(\boldsymbol{m}) = 1. \tag{2.8}$$

The given SGBE has no explicit solution, and this fact complicates the solution of the considered problem for large state space dimensions (2.1).

In this regard, we propose another approach based on the use of the fact that the QoS metrics (2.5) and (2.6) are defined in terms of the probabilities of the merged states S_n, $n = 0, 1, \ldots, N$. Since the sets S_n, $n = 0, 1, \ldots, N$ define some splitting of the state space (2.1), then the desired QoS metrics can be calculated using the probabilities of merged states. Indeed, it is clear that the probabilities of the merged states are defined as follows:

$$\pi(n) = \sum_{\boldsymbol{m} \in S_n} p(\boldsymbol{m}), \quad n = 0, 1, \ldots, N. \tag{2.9}$$

It is obvious that (see normalizing condition (2.8))

$$\sum_{n=0}^{N} \pi(n) = 1. \tag{2.10}$$

Consequently, taking into account (2.5), (2.6), (2.9), and (2.10), one finds that

$$\text{PB}_i = \sum_{n=0}^{N} (1 - \alpha_i(n)) \pi(n), \quad i = 1, \ldots, K, \tag{2.11}$$

$$N_{\text{av}} = \sum_{n=1}^{N} n\pi(n). \tag{2.12}$$

Thus, without determining the stationary distribution of the original (initial) model, one can calculate the QoS metrics (2.5) and (2.6), if it is possible to determine the values of the probabilities of merged states $\pi(n)$, $n=0,1,\ldots,N$. With the help of the following statement, one can solve this problem.

Proposition 2.1 If $\mu_i=\mu_j$, $i,j=1,\ldots,K$, then QoS metrics (2.5) and (2.6) are defined as follows:

$$\text{PB}_i = \left(\sum_{n=0}^{N}(1-\alpha_i(n))g_n\right)\Big/\left(\sum_{n=0}^{N} g_n\right), \quad i=1,\ldots,K, \tag{2.13}$$

$$N_{\text{av}} = \left(\sum_{n=1}^{N} n g_n\right)\Big/\left(\sum_{n=0}^{N} g_n\right). \tag{2.14}$$

Henceforward, the following notations are used: $v_i=\lambda_i/\mu_i$, $i=1,\ldots,K$;

$$\widetilde{v}_i(n-i) = \sum_{j\in A(i)} v_j\alpha_j(n-i), \tag{2.15}$$

where $A(i)=\{j: j\text{-calls demand } i \text{ channels}\}$, $i=1,\ldots,N$;

$$g_n = \begin{cases} 1, & n=0, \\ \dfrac{1}{n}\displaystyle\sum_{i=1}^{n} i\widetilde{v}_i(n-i)g_{n-i}, & n=1,\ldots,N. \end{cases} \tag{2.16}$$

To prove Proposition 2.1, let us first prove the following fact.

Proposition 2.2 If $\mu_i=\mu_j$, $i,j=1,\ldots,K$, then probabilities of merged states are defined as

$$\pi(n) = g_n\pi(0), \quad n=0,1,\ldots,N, \tag{2.17}$$

where $\pi(0)=\left(\sum_{n=0}^{N} g_n\right)^{-1}$.

Note 2.2 In special case, if $\alpha_i(n)=\alpha_j(n)$ for $b_i=b_j$, from Eqs. (2.15) and (2.16) one finds that the parameters g_n, $n=0,1,\ldots,N$ are determined as follows:

$$g_n = \begin{cases} 1, & n = 0 \\ \frac{1}{n}\sum_{i=1}^{n} i\widetilde{v}_\iota \alpha_i(n-i) g_{n-i}, & n = 1, \ldots, N, \end{cases}$$

where $\widetilde{v}_\iota = \begin{cases} \sum_{j \in A(i)} v_j & \text{if}\, A(i) \neq \varnothing, \\ 0 & \text{if}\ A(i) = \varnothing. \end{cases}$

Proposition 2.2 is a direct consequence of a following one.

Proposition 2.3 If $\mu_i = \mu_j$, $i, j = 1, \ldots, K$, then the following equalities hold true:

$$\sum_{i=1}^{K} v_i b_i \alpha_i (n - b_i) \pi(n - b_i) = n\pi(n), \quad n = 1, \ldots, N, \tag{2.18}$$

where $\pi(x) = 0$ for $x < 0$.

Proof of Proposition 2.3 For simplicity, we present the proof of this fact for a single-rate model, i.e., for a model in which $b_i = 1$ for all $i = 1, \ldots, K$. Generalization for a multi-rate model is straightforward.

Let us use the scheme proposed in [13]. Taking into account relationship (2.4), one obtains that the SGBE for the states $\boldsymbol{m} \in S_{n-1}$ has the following form:

$$\left(\sum_{i=1}^{K} \lambda_i \alpha_i (n-1) + \sum_{i=1}^{K} m_i \mu_i\right) p(\boldsymbol{m}) = \sum_{i=1}^{K} \lambda_i \alpha_i (n-2) p(\boldsymbol{m} - \boldsymbol{e}_i)$$

$$+ \sum_{i=1}^{K} (m_i + 1) \mu_i p(\boldsymbol{m} + \boldsymbol{e}_i). \tag{2.19}$$

For simplicity, it is assumed that the states $\boldsymbol{m}$, $\boldsymbol{m} - \boldsymbol{e}_i$, $\boldsymbol{m} + \boldsymbol{e}_i$ participating in Eq. (2.19) are in state space (2.1); otherwise the corresponding members are zeroed.

Summing both sides of Eq. (2.19) over all possible $\boldsymbol{m} \in S_{n-1}$, after collecting similar terms and taking into account structure of SGBE, one gets

$$\sum_{i=1}^{K} \lambda_i \alpha_i (n-1) \sum_{\boldsymbol{m} \in S_{n-1}} p(\boldsymbol{m}) = \sum_{\boldsymbol{m} \in S_n} m_i \mu_i p(\boldsymbol{m}). \tag{2.20}$$

In latter transformations, while rearranging the terms in the sum, the two facts essentially have been taken into account: relationship (2.9) as well as the following fact: for all states $\boldsymbol{m} \in S_n$, $n = 1, \ldots, N$ the value $\sum_{j=1}^{K} v_j \alpha_j(n)$ is the same.

Taking into account Eq. (2.9), Eq. (2.20) might be rewritten as follows:

$$\pi(n-1)\sum_{i=1}^{K}\lambda_i\alpha_i(n-1)=\sum_{m\in S_n} m_i\mu_i p(\boldsymbol{m}). \qquad (2.21)$$

From Eq. (2.21) for $\mu_i=\mu_j$, $i,j=1,\ldots,K$, we have

$$\pi(n-1)\sum_{i=1}^{K}v_i\alpha_i(n-1)=\sum_{m\in S_n} m_i p(\boldsymbol{m}). \qquad (2.22)$$

The right side of Eq. (2.22) can be represented as follows:

$$\sum_{m\in S_n} m_i p(\boldsymbol{m})=\sum_{m\in S_n} m_i\frac{p(\boldsymbol{m})}{\pi(n)}\pi(n). \qquad (2.23)$$

From the definition of the conditional probability, one has

$$P(\boldsymbol{m}|n)=P\left(\boldsymbol{m}\middle|\sum_{i=1}^{K}m_i=n\right)=\begin{cases}\dfrac{p(\boldsymbol{m})}{\pi(n)} & \text{if } \boldsymbol{m}\in S_n\\ 0 & \text{in other cases,}\end{cases} \qquad (2.24)$$

where $P(\cdot\,|\,\cdot)$ is a sign of conditional probability.

Then from Eq. (2.23), taking into account Eq. (2.24), one obtains

$$\begin{aligned}\sum_{i=1}^{K}\sum_{m\in S_n} m_i p(\boldsymbol{m})&=\sum_{i=1}^{K}\left(\sum_{m\in S_n} m_i P(\boldsymbol{m}|n)\right)\pi(n)\\&=\sum_{i=1}^{K}E(m_i|n)\pi(n)=E\left(\sum_{i=1}^{K}m_i|n\right)\pi(n)=n\pi(n),\end{aligned} \qquad (2.25)$$

where $E(\cdot\,|\,\cdot)$ is a sign of conditional expectation.

Consequently, taking into account (2.22) and (2.25), one concludes that relationships (2.18) are valid for single-rate model. As it has been noted above, the generalization of this proof for a multi-rate model is straightforward.

Now one can prove *Proposition* 2.2. Indeed, after some algebraic transformations one finds that the system of equations (2.18), taking into account the normalization condition (2.10), has the following augmented matrix:

$$\begin{pmatrix} \tilde{v}_1(0) & -1 & 0 & \dots & 0 & 0 & 0 \\ 2\tilde{v}_2(0) & \tilde{v}_1(1) & -2 & \dots & 0 & 0 & 0 \\ \cdot & \cdot & \cdot & \cdot & \cdot & \cdot & \cdot \\ N\tilde{v}_N(0) & (N-1)\tilde{v}_{N-1}(1) & (N-2)\tilde{v}_{N-2}(2) & \dots & \tilde{v}_1(N-1) & -N & 0 \\ 1 & 1 & 1 & \dots & 1 & 1 & 1 \end{pmatrix}$$

Hence, one finds that the stationary probabilities of merged states, while using randomized access scheme, are determined from Eq. (2.17).

Consequently, taking into account (2.11) and (2.12), one finds that QoS metrics of model (2.5) and (2.6) are calculated from relations (2.13) and (2.14). In other words, *Proposition* 2.1 is proved.

An important advantage of this algorithm is that its computational complexity does not depend on the total number of types of calls (i.e., on K) and is estimated as $O(N)$. Such invariance is achieved through merging of flows by the number of required channels (see the definition of the sets $A(n)$, $n = 1, \dots, N$).

Note that in special cases we exactly obtain results from the above-indicated well-known access schemes based on CS, CSE, and TR schemes (see *Propositions* 1.3, 1.4, and 1.5 in Sect. 1.1).

To calculate the QoS metrics of the system, the computational procedure, described above, can be used even in cases where the service intensities of heterogeneous calls are unessentially different from one another. And in the cases, where the service intensities of heterogeneous calls are essentially different, one can use different schemes of "unification" ("averaging") of their values. So, from a practical standpoint, the use of the following three general values is most interesting:

(1) $\mu = \max\{\mu_1, \dots, \mu_K\}$; (2) $\mu = \min\{\mu_1, \dots, \mu_K\}$; (3) $\mu = \frac{1}{\Lambda}\sum_{i=1}^{K} v_i$, where $\Lambda = \sum_{i=1}^{K}\lambda_i$.

Note that for every "averaging" scheme (this and others), the accuracy of the used approximations can be studied numerically, since analytical solution does not exist. For the models of small dimension, the exact solution can be found from SGBE.

2.1.1 Model of Integral Wireless Network with Multi-parametric Access Scheme

The multi-rate queueing model in the problems of calculating the QoS metrics of the integral wireless communication network of a cellular structure was used in [25]. In mentioned work, four types of calls are considered: handover voice calls (hv-calls), new voice calls (ov-calls), handover data calls (hd-calls), and new data calls (od-calls). The network uses a fixed channel allocation scheme (FCA scheme), and every cell has $N > 1$ radio channels. The intensity of the x-calls is λ_x, $x \in$ {hv, ov, hd, od}.

To service one voice call (v-call), it is required only one free channel, and one nonelastic data call (d-call) requires $b > 1$ channels simultaneously. The distribution functions of channels holding time by heterogeneous calls are exponential, the average intensity of processing one v-call (new or handover) is μ_v, and the corresponding parameter for d-calls (new or handover) is μ_d.

In the system the following multi-parametric access scheme is used (see [25]). For defining the success scheme, three parameters N_1, N_2, and N_3 are introduced. It is assumed that parameters N_1 and N_2 are multiples of b. These parameters satisfy the inequality $0 < N_1 \leq N_2 \leq N_3 \leq N$. The proposed access scheme is determined by the following rules of receiving heterogeneous calls:

- If upon arrival of an od-call the number of busy channels is no more than $N_1 - b$, it is served; otherwise, it is rejected.
- If upon arrival of an hd-call the number of busy channels is no more than $N_2 - b$, it is served; otherwise, it is rejected.
- If upon arrival of an ov-call the number of busy channels is less than N_3, it is served; otherwise, it is rejected.
- If upon arrival of an hv-call there is at least one free channel, it is served; otherwise, it is rejected.

To calculate the QoS metrics of the pointed model in [25], a recursive method is developed. This method faces the well-known computational difficulties for models of a large dimension. The approximate method of solving this problem for certain ratios of loads of heterogeneous traffics is developed in [16]. Here we develop an accurate and computationally efficient method to solving this problem [17, 23].

It is easy to see that the given model of an integral network with the multi-parametric access scheme is a special case of the model, studied in Sect. 2.1, with a randomized access scheme. Indeed, we obtain the given model if in the studied above model one will set $K = 4$ and the parameters $\alpha_i(n)$ will be determined as follows:

$$\alpha_{od} = \begin{cases} 1 & \text{if } n \leq N_1 - b, \\ 0 & \text{in other cases;} \end{cases} \tag{2.26}$$

$$\alpha_{hd} = \begin{cases} 1 & \text{if } n \leq N_2 - b, \\ 0 & \text{in other cases;} \end{cases} \tag{2.27}$$

$$\alpha_{ov} = \begin{cases} 1 & \text{if } n < N_3, \\ 0 & \text{in other cases;} \end{cases} \tag{2.28}$$

$$\alpha_{hv} = \begin{cases} 1 & \text{if } n < N, \\ 0 & \text{in other cases.} \end{cases} \tag{2.29}$$

In view of Eqs. (2.26)–(2.29), one finds that the sets $A(i)$, $i = 1, \ldots, N$ (see formula (2.15)) are defined as follows:

$$A(i) = \begin{cases} \{\text{ov}, \text{hv}\} & \text{if } i = 1, \\ \{\text{od}, \text{hd}\} & \text{if } i = b, \\ \varnothing & \text{in other cases.} \end{cases}$$

After certain transformations from Eqs. (2.15), (2.16), one gets that in this model for $\mu_v = \mu_d$ the parameters g_n, $n = 0, 1, \ldots, N$ are determined from the following simple formulas:

$$g_0 = 1,$$

$$g_n = \frac{1}{n}((v_v I(n-1 < N_3) + v_{hv} I(n-1 \geq N_3))g_{n-1} + b(v_d I(n \leq N_1)$$

$$+ v_{hd} I(N_1 < n \leq N_2))g_{n-b}),$$

where $n = 1, \ldots, N$ and $g_x = 0$ if $x < 0$. Here the following notation is taken: $v_{ov} = \lambda_{ov}/\mu_v$, $v_{hv} = \lambda_{hv}/\mu_v$, $v_v = v_{ov} + v_{hv}$; $v_{od} = \lambda_{od}/\mu_d$, $v_{hd} = \lambda_{hd}/\mu_d$, $v_d = v_{od} + v_{hd}$.

Consequently, the desired QoS metrics are calculated as follows:

$$\text{PB}_{\text{hv}} = \pi(N); \text{PB}_{\text{ov}} = \sum_{n=N_3}^{N} \pi(n); \text{PB}_{\text{od}}$$

$$= \sum_{n=N_1-b+1}^{N} \pi(n); \ \text{PB}_{\text{hd}} = \sum_{n=N_2-b+1}^{N} \pi(n). \tag{2.30}$$

In other words, the computation of the QoS metrics of the given model by formulas (2.30) does not offer difficulties for models of any dimension, and it is much easier than the known algorithms [16, 25].

In the case $\mu_v \neq \mu_d$, as it is indicated above, one can use different schemes of approximate solution of the problem.

Note that problems of obtaining prescribed QoS level for heterogeneous calls are of definite scientific and practical interest. In this case, some adjustable parameters should exist for the solution of such problems. In this connection, note that, in networks with FCA schemes, only threshold parameters of the CAC scheme can be controlled since the control of load parameters is a rather complicated and sometimes even an unsolvable problem from the practical viewpoint.

Here a problem of finding the set of values of threshold parameters of the described above CAC scheme is considered for which a prescribed QoS level for heterogeneous calls is satisfied. We call this set (if it is not empty) the set of efficient values (SEVs) of threshold parameters.

For the models being investigated, there exist great possibilities of solution of these problems since there are three degrees of freedom (i.e., the thresholds N_1, N_2, and N_3) in them. Hence, various statements of problems of finding the set of efficient values of threshold parameters are possible.

A verbal definition of the problem being considered is as follows. In an FCA scheme under fixed loads, upper bounds are prescribed for possible values of loss probabilities of heterogeneous calls. It is required to find values of threshold parameters N_1, N_2, and N_3 that satisfy the prescribed constraints.

For small values of N, the solution of this problem can be found by a simple exhaustive search for all possible combinations of parameters N_1, N_2, and N_3. However, this approach becomes inefficient with the growth in N and sometimes is simply impossible. Therefore, an algorithmic approach is proposed below to the solution of the mentioned problem without using an exhaustive search for variants.

For simplicity, assume that new and handover calls are not distinguished in a data traffic, i.e., assume that $N_1 = N_2$. Then, according to relationship (2.30), we have $PB_{od} = PB_{hd}$.

We denote $PB_d = PB_{od} = PB_{hd}$. Then the problem is mathematically written as follows: it is required to find pairs (N_2, N_3) where $N_2 \leq N_3$, for which the following constraints are satisfied:

$$PB_{hv} \leq \varepsilon_{hv}, \tag{2.31}$$

$$PB_{ov} \leq \varepsilon_{ov}, \tag{2.32}$$

$$PB_d \leq \varepsilon_d, \tag{2.33}$$

where ε_{hv}, ε_{ov}, and ε_d are given values.

A possible algorithm for solution of problem (2.31)–(2.33) using monotonic property of QoS metrics being investigated is presented below.

The main idea of such an iterative algorithm is as follows: for each fixed value of the parameter N_3, the search for the set of efficient values is performed due to the choice of the corresponding values of the parameter N_2. For convenience, this argument is shown in the notation of these functions.

For generality, we consider the kth iteration, $k = 1, 2, \ldots, N$.

Step 1 Set $N_3 = k$ and check the following conditions:

$$PB_{hv}(1) \leq \varepsilon_{hv}, \tag{2.34}$$

$$PB_{ov}(1) \leq \varepsilon_{ov}, \tag{2.35}$$

$$PB_d(N_3) \leq \varepsilon_d. \tag{2.36}$$

If all conditions (2.34)–(2.36) are satisfied, go to the next step. Otherwise, for the prescribed value of N_3, the problem has no solution.

Note 2.3 Since the function PB_{hv} does not decrease with respect to the parameter N_3, the nonfulfillment of condition (2.34) for a definite value of N_3 implies its nonfulfillment for all $k > N_3$. Allowance for this fact considerably accelerates the operation of the algorithm.

Step 2 Solve the following problem:

$$\underline{N_2} = \arg\min_{N_2 \in [1,N_3]} \{\mathrm{PB_d}(N_2) \leq \varepsilon_\mathrm{d}\}. \tag{2.37}$$

Step 3 If $\mathrm{PB_{hv}}\left(\underline{N_2}\right) \leq \varepsilon_\mathrm{hv}$ and $\mathrm{PB_{ov}}\left(\underline{N_2}\right) \leq \varepsilon_\mathrm{ov}$, then go to the next step. Otherwise, for this value of N_3, the problem has no solution.

Step 4 Simultaneously solve the following problems:

$$N_2^\mathrm{hv} = \arg\max_{N_2 \in [\underline{N_2},N_3]} \{\mathrm{PB_{hv}}(N_2) \leq \varepsilon_\mathrm{hv}\}, \tag{2.38}$$

$$N_2^\mathrm{ov} = \arg\max_{N_2 \in [\underline{N_2},N_3]} \{\mathrm{PB_{ov}}(N_2) \leq \varepsilon_\mathrm{ov}\}. \tag{2.39}$$

Step 5 Determine the sought-for interval of appropriate values of N_2 for a given value of N_3 as $\left[\underline{N_2}, \overline{N_2}\right]$ where $\overline{N_2} = \min\left(N_2^\mathrm{hv}, N_2^\mathrm{ov}\right)$.

Step 6 If $N_3 < N$, then set $N_3 = N_3 + 1$ and go to step 1. Otherwise, terminate the algorithm.

Note 2.4 Based on monotonic property of the functions being investigated, the dichotomy (binary search) method can be used for the solution of problems (2.37)–(2.39).

Hence, for each fixed value of the threshold N_3, the set of admissible values of N_2 is found (if they exist), and the set of efficient values of threshold parameters is found by uniting all the solutions obtained.

Numerical experiments were performed using the developed algorithm. For a sample model, the following initial data for test problems (2.31)–(2.33) were used: $N = 50$, $\nu_\mathrm{ov} = 8/9$, $\nu_\mathrm{hv} = 1/3$, $\nu_\mathrm{od} = 1/2$, $\nu_\mathrm{hd} = 1/4$. The corresponding SEVs for the problem under various constraints on the values of loss probabilities of heterogeneous calls are shown in Table 2.1. Here, the Cartesian product $[a, b] \times [c, d]$ means that $N_2 \in [a, b]$ and $N_3 \in [c, d]$.

As is obvious from Table 2.1, the weakening of requirements on the QoS metrics of d-calls leads to an extension of SEVs owing to the decrease in inefficient values of the parameter N_2 (see rows 1–4 in Table 2.1). This would be expected since the loss probability of d-calls decreases with increasing the parameter N_2. In this case, an SEV is rather smoothly extended with respect to the change in the upper bound of the loss probability of d-calls (i.e., ε_d). It should also note that, for a fixed value of ε_d, an SEV retains its form for rather wide range of varying the other bounds ε_hv and ε_ov (see rows 5–8 in Table 2.1).

In practice, loads of heterogeneous traffics are changed in time. Therefore, problems of investigating the sensitivity of efficient values of threshold parameters with respect to a change in loads are topical questions. In this connection, we note that any analytical investigation of this problem is impossible in principle; it can be investigated only by means of numerical experiments. In particular, performed numerical experiments show that efficient values of threshold parameters of

Table 2.1 Results of solution of problem (2.31)–(2.33)

Parameter values			
ε_{hv}	ε_{ov}	ε_{d}	SEV
10^{-4}	10^{-5}	10^{-6}	[12,30] × [31,50]
10^{-4}	10^{-5}	10^{-5}	[10, 30] × [31,50]
10^{-4}	10^{-5}	10^{-4}	[9, 30] × [31,50]
10^{-4}	10^{-5}	10^{-3}	[8, 30] × [31,50]
10^{-4}	10^{-4}	10^{-3}	[8, 30] × [31,50]
10^{-4}	10^{-3}	10^{-3}	[8, 30] × [31,50]
10^{-4}	10^{-2}	10^{-3}	[8, 30] × [31,50]
10^{-2}	10^{-4}	10^{-3}	[8, 30] × [31,50]

problem (2.31)–(2.33) are preserved within a sufficiently wide load variation interval. This is explained by a rather smooth change in the QoS metrics being investigated with respect to loads of heterogeneous traffics.

2.1.2 Numerical Results

Let us consider the results of numerical experiments for the model of multi-rate queue with three types of traffics. The appropriate algorithm to calculate the QoS metrics is quite simple and allows us to study their behavior in all ranges of changing the values of the load and structural parameters of the system. To keep it brief only dependency of QoS metrics on the number of channels is shown in two schemes of determining the access probabilities of heterogeneous calls. In both schemes the bandwidth and load parameters are fixed and chosen as follows: $b_1=2$, $b_2=5$, $b_3=8$; $v_1=0.03$ Erl, $v_2=0.02$ Erl, $v_3=0.01$ Erl.

In the first scheme it is assumed that $\alpha_i(j)=b_i/(j+b_i)$, and in the second one, $\alpha_i(j)=(j+1)/(j+b_i)$, $i=1,2,3$. In other words, in the first scheme access probabilities of heterogeneous calls are determined by decreasing function, while in the second one the indicated parameters are defined by increasing function with respect to the number of busy channels. Besides these properties, the introduced access probabilities have the following properties: for the first scheme $\alpha_1(j)<\alpha_2(j)<\alpha_3(j)$; for the second scheme $\alpha_1(j)>\alpha_2(j)>\alpha_3(j)$ for any j, $j=1,2,\ldots,N$. It means that in the first scheme for the given number of busy channels, access probabilities are determined by increasing function with respect to their bandwidth, while in the second scheme we have inverse situation.

Here our goal is comparison of QoS metrics of the system under different access schemes. Corresponding results are summarized in Figs. 2.1, 2.2, and 2.3 where labels 1 and 2 denote loss probabilities for the first scheme and second scheme, respectively. Their analysis enables us to make the following conclusions. First of all, note that all the QoS metrics under study are decreasing functions with respect to the total number of channels. They completely confirmed all theoretical expectations. However, unlike the function PB_1 the rates of change of the functions PB_2

Fig. 2.1 Blocking probability of calls of the first type versus N

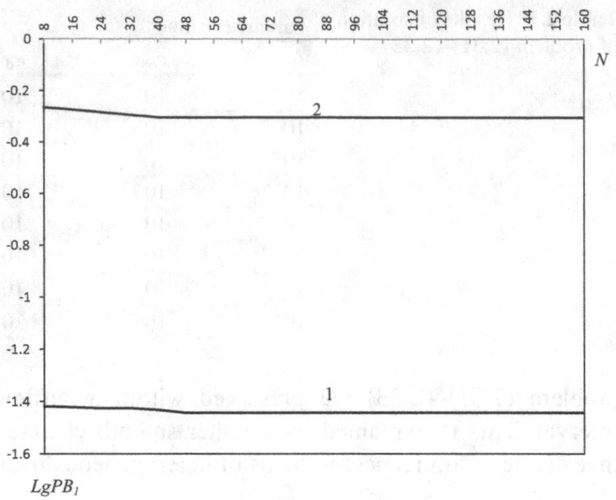

Fig. 2.2 Blocking probability of calls of the second type versus N

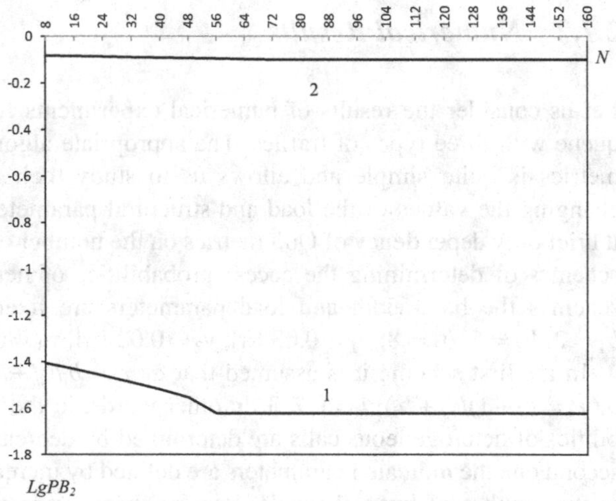

and PB_3 in first scheme are sufficiently high for small values of channels, i.e., for $N \geq 56$ both functions PB_2 and PB_3 become almost constant.

It is worth noting that for given initial data the first scheme is preferable. However, quite probably, for other values of initial data of the QoS metrics (either all or some of them), the second scheme will be better than the first one. Note that finding the optimal (in known sense) values of access probabilities is not a trivial problem especially for large-size models with many types of heterogeneous calls. However, note that for solving such kind of problems, the methods of Markov decision processes are useful.

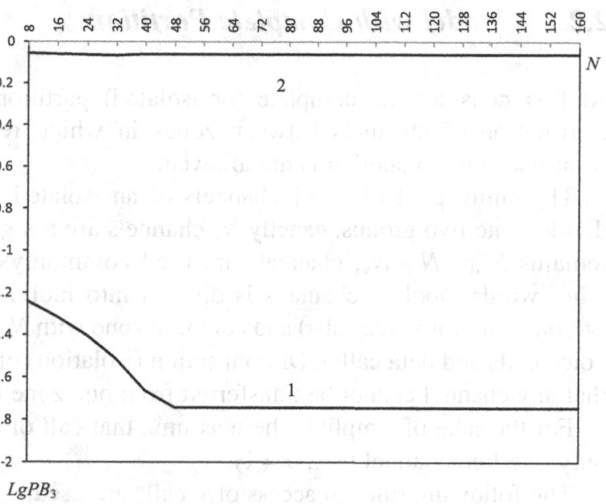

Fig. 2.3 Blocking probability of calls of the third type versus N

2.2 Models of Integral Cellular Networks with Partition of Channels

In majority of known CAC schemes all channels of a cell are equally accessible to a call of any type. However, to reduce the possibility of occurrence of conflict situations using the appropriate schemes of partition of the entire pool of channels between heterogeneous calls is effective also. The analysis of the accessible literature has shown that models of integral cellular networks with such kind of access schemes are insufficiently investigated. Note that the isolated (rigid) partition of channels not always is effective [9], so other partition schemes are required. Thereupon note that non-isolated schemes of partition of channels in network with single traffic (networks of the second generation) have been offered in [20] and in [27], Chap. 1. Feature of these schemes consists that in them partition of channels is not rigid, i.e., the scheme of virtual partition of channels (virtual partitioning, VP) is used.

Here two multi-parametric schemes to partition of entire pool of channels between heterogeneous calls in models of integral wireless networks with voice and data calls (see Sect. 2.1.1) are proposed. Exact algorithms to calculate the QoS metrics of heterogeneous calls under given partition schemes are developed. Results related to determining the efficient partition scheme are carried out.

2.2.1 Model with Complete Partition

At first consider the complete (or isolated) partition scheme (CP scheme) for distribution of channels between zones in which reassignment of the channel from one zone to another is not allowed.

The entire pool of $N>1$ channels of an isolated cell of integral networks is divided into two groups: exactly N_v channels are assigned for voice calls only and remains $N_{vd}=N-N_v$; channels are used commonly by voice and data calls. In other words, pool of channels is divided into individual zone with N_v channels (v-zone only for voice calls) and common zone with N_{vd} channels (vd-zone for both voice calls and data calls). Disconnection (isolation) of division of channels means that any channel cannot be transferred from one zone to another.

For the sake of simplicity here assume that call of any type to service required only one free channel (i.e., $b=1$).

The following rules to access of v-calls are used:

- If upon arrival of an ov-call there is at least one free channel in v-zone, this call seizes one of them; otherwise this call is rejected.
- If upon arrival of an hv-call there is at least one free channel in v-zone, this call seizes one of them; otherwise free channel is searched in vd-zone. At that there is a limit to the number of hv-calls in vd-zone, i.e., an hv-call is accepted to vd-zone only if the number of hv-calls in this zone is less than R_{hv}, $1 \le R_{hv} \le N_{vd}$; otherwise it is rejected.

Note that channel holding time of hv-calls in vd-zone has exponential distribution with the same average $1/\mu_v$.

Accesses of d-calls are controlled by the following rules:

- If upon arrival of an hd-call there is at least one free channel in vd-zone, this call seizes one of them; otherwise this call is rejected.
- Arrived od-call is accepted to vd-zone only if the number of d-calls in this zone is less than R_{od}, $1 \le R_{od} \le N_{vd}$; otherwise it is rejected.

Consider the problem of finding the main QoS metrics (i.e., loss probabilities of heterogeneous calls) of the network under given partition scheme of the channels.

From the description of the proposed CP scheme, we conclude that the loss probability of new voice calls is easily defined as loss probability in a classical Erlang's model $M/M/N_v/N_v$ with load ν_v Erl, where $\nu_v=(\lambda_{ov}+\lambda_{hv})/\mu_v$. In other words, to calculate this QoS metrics the well-known Erlang's B-formula might be used:

$$PB_{ov} = E_B(\nu_v, N_v) \tag{2.40}$$

where $E_B(\nu,n)=(\nu^n/n\,!)/(\sum_{i=0}^{n}(\nu^i/i\,!))$.

However, the loss probability of hv-calls cannot be defined by means of the formula (2.40) since the hv-calls not accepted in the v-zone under certain conditions are transferred to vd-zone. Thus, intensity of hv-calls to vd-zone $\left(\widetilde{\lambda}_{hv}\right)$ is determined as $\widetilde{\lambda}_{hv} = \lambda_{hv} PB_{ov}$.

Therefore, to calculate the remaining three QoS metrics, it is required to study the multi-flow Erlang's model $M/M/N_{vd}/N_{vd}$ with three types of calls, i.e., hv-calls (with intensity $\widetilde{\lambda}_{hv}$), od-calls (with intensity λ_{od}), and hd-calls (with intensity λ_{hd}). Since the channel holding times of heterogeneous calls differ from each other, the state of the mentioned model is described by 2-D vector $\boldsymbol{n} = (n_d, n_v)$, where n_d (respectively, n_v) is the total number of data (respectively, handover voice calls) calls in the channels. Then the state space of the corresponding 2-D MC describing this model is defined thus

$$S = \{\boldsymbol{n} : \ n_d = 0, 1, \ldots, N_{vd}; \ n_v = 0, 1, \ldots, R_{hv}; \ n_d + n_v \le N_{vd}\}.$$

Taking into account the proposed CP scheme for heterogeneous calls, we conclude that the nonnegative elements of the Q-matrix of the appropriate 2-D MC in this model are determined as follows (see Fig. 2.4):

$$q(\boldsymbol{n}, \boldsymbol{n}') = \begin{cases} \lambda_d & \text{if } n_d < R_{od}, \ \boldsymbol{n}' = \boldsymbol{n} + \boldsymbol{e}_1, \\ \lambda_{hd} & \text{if } n_d \ge R_{od}, \ \boldsymbol{n}' = \boldsymbol{n} + \boldsymbol{e}_1, \\ \widetilde{\lambda}_{hv} & \text{if } n_v < R_{hv}, \ \boldsymbol{n}' = \boldsymbol{n} + \boldsymbol{e}_2, \\ n_d \mu_d & \text{if } \boldsymbol{n}' = \boldsymbol{n} - \boldsymbol{e}_1, \\ n_v \mu_v & \text{if } \boldsymbol{n}' = \boldsymbol{n} - \boldsymbol{e}_2, \\ 0 & \text{in other cases}, \end{cases} \tag{2.41}$$

where $\lambda_d = \lambda_{od} + \lambda_{hd}$, $\boldsymbol{e}_1 = (1, 0)$, $\boldsymbol{e}_2 = (0, 1)$.

It is easy to show that all states of this 2-DMC are communicating, so in this chain stationary mode exists. Let $p(\boldsymbol{n})$ denote the stationary probability of state $\boldsymbol{n} \in S$.

Desired QoS metrics of the proposed CP scheme are determined via marginal distribution of the above-indicated 2-D MC. Indeed, in this scheme losses of hv-calls occur in the following cases: (a) upon arrival of hv-call, the number of calls of this type in the system is equal R_{hv} regardless of the number of busy channels, and (b) upon arrival of hv-calls, all channels are busy. Therefore by using PASTA theorem, we obtain

$$PB_{hv} = \sum_{\boldsymbol{n} \in S} p(\boldsymbol{n}) (\delta(n_v, R_{hv})(1 - \delta(n_d + n_v, N_{vd})) + (1 - \delta(n_v, R_{hv}))\delta(n_d + n_v, N_{vd})). \tag{2.42}$$

Arguing similarly, we find that loss probabilities of od-calls (PB_{od}) and hd-calls (PB_{hd}) are determined as follows:

Fig. 2.4 State transition diagram of the model with CP scheme for partition of channels

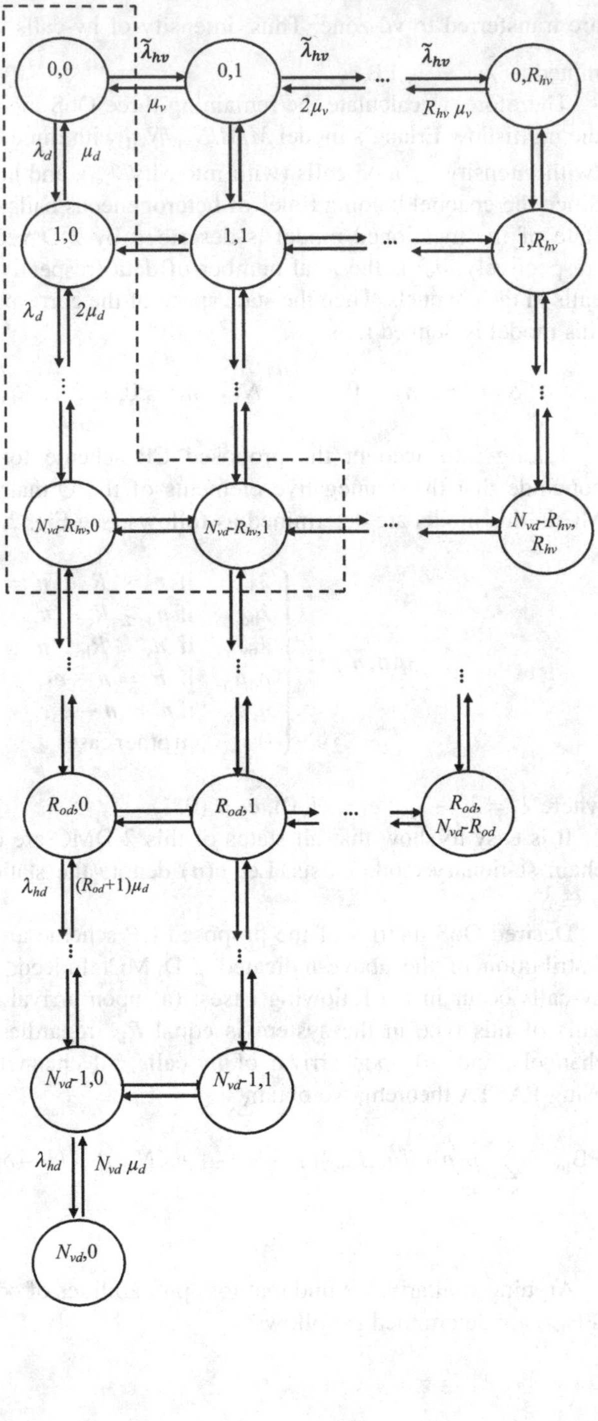

$$\mathrm{PB_{od}} = \sum_{\boldsymbol{n}\in S} p(\boldsymbol{n}) I(n_\mathrm{d} \geq R_\mathrm{od}); \tag{2.43}$$

$$\mathrm{PB_{hd}} = \sum_{\boldsymbol{n}\in S} p(\boldsymbol{n}) (n_\mathrm{d} + n_\mathrm{v}, N_\mathrm{vd}). \tag{2.44}$$

System of global balance equations (SGBE) for stationary probabilities is constructed by using relationship (2.41) (we left it to reader). However stationary probabilities can be determined analytically without the numerical solution of the indicated SGBE which for real system has a large dimension.

Proposition 2.4 Stationary distribution of the system at use CP scheme has the following multiplicative form:

$$p(i,j) = \begin{cases} \dfrac{\nu_\mathrm{d}^i}{i!}\dfrac{\widetilde{\nu}_\mathrm{hv}^j}{j!} p(0,0) & \text{if } 0 \leq i \leq R_\mathrm{od},\ 0 \leq j \leq \min(R_\mathrm{hv}, N_\mathrm{vd} - i), \\ \left(\dfrac{\nu_\mathrm{d}}{\nu_\mathrm{hd}}\right)^{R_\mathrm{od}} \dfrac{\nu_\mathrm{hd}^i}{i!}\dfrac{\widetilde{\nu}_\mathrm{hv}^j}{j!} p(0,0) & \text{if } R_\mathrm{od}+1 \leq i \leq N_\mathrm{vd},\ 0 \leq j \leq \min(R_\mathrm{hv}, N_\mathrm{vd} - i), \end{cases} \tag{2.45}$$

where $p(0,0)$ is determined from the normalizing condition, i.e., $\sum_{\boldsymbol{n}\in S} p(\boldsymbol{n}) = 1$ and $\widetilde{\nu}_\mathrm{hv} = \widetilde{\lambda}_\mathrm{hv}/\mu_\mathrm{v}$.

Proof of this fact is based on Kolmogorov's theorem about reversibility of 2-D MC [18]. Indeed, it is easily shown that there is no circulation between states $\boldsymbol{n}, \boldsymbol{n}+\boldsymbol{e}_1, \boldsymbol{n}+\boldsymbol{e}_2, \boldsymbol{n}+\boldsymbol{e}_1+\boldsymbol{e}_2$ of the state diagram of the underlying 2-D MC, i.e., there is a general solution of the system of local balance equations (SLBE) for state probabilities. Thus by choosing the path (0, 0), (1, 0), ..., $(i, 0)$, $(i, 1)$, ..., (i, j) from state (0, 0) to state (i, j), we find that multiplicative solution (2.45) is holding (see Fig. 2.4, area in a dashed line). Note that in this proof scheme it is required to take into account two cases which are indicated in the right side of formula (2.45).

Finally, after calculating the state probabilities, the QoS metrics (2.42)–(2.44) are determined from the following explicit formulas:

$$\mathrm{PB_{hv}} = \sum_{i=0}^{N_\mathrm{vd}-R_\mathrm{hv}} p(i, R_\mathrm{hv}) + \sum_{i=N_\mathrm{vd}-R_\mathrm{hv}+1}^{N_\mathrm{vd}} p(i, N_\mathrm{vd} - i), \tag{2.46}$$

$$\begin{aligned} \mathrm{PB_{od}} = {} & I(R_\mathrm{od} \leq N_\mathrm{vd} - R_\mathrm{hv}) \sum_{i=R_\mathrm{od}}^{N_\mathrm{vd}} \sum_{j=0}^{\min(R_\mathrm{hv}, N_\mathrm{vd}-i)} p(i,j) \\ & + I(R_\mathrm{od} > N_\mathrm{vd} - R_\mathrm{hv}) \left(\sum_{i=N_\mathrm{vd}-R_\mathrm{hv}}^{R_\mathrm{od}-1} p(i, N_\mathrm{vd} - i) + \sum_{i=R_\mathrm{od}}^{N_\mathrm{vd}} \sum_{j=0}^{N_\mathrm{vd}-i} p(i,j) \right), \end{aligned} \tag{2.47}$$

$$PB_{hd} = \sum_{i=N_{vd}-R_{hv}}^{N_{vd}} p(i, N_{vd} - i). \tag{2.48}$$

2.2.2 *Model with Virtual Partition*

Now consider similar model with virtual partition (VP scheme) of channels between two zones. The basic difference of the given scheme from the previous one consists in the following: upon completion of servicing of a v-call in v-zone, the relinquished channel is transferred to the vd-zone if there is a v-call present here, while the channel in the vd-zone that has servicing v-call is switched to the v-zone. In other words, partition is a virtual one, and this procedure is similar to channel reallocation scheme.

Note that at use VP scheme the loss probability of new voice calls cannot be calculated simply from classical Erlang's B-formula (2.40). It is explained by the fact that in this scheme reallocation of channels is allowed.

As abovementioned scheme, here the state of the model is described by 2-D vector $\boldsymbol{n} = (n_d, n_v)$ also, where n_d (respectively, n_v) is the total number of data (respectively, handover voice calls) calls in the channels. However, the state space of the corresponding 2-D MC is defined as follows:

$$S = \{\boldsymbol{n} : n_d = 0, 1, \ldots, N_{vd};\ n_v = 0, 1, \ldots, N_v + R_{hv};\ n_d + n_v \le N\}.$$

In VP scheme the nonnegative elements of the Q-matrix of the appropriate 2-D MC is determined as follows (see Fig. 2.5):

$$q(\boldsymbol{n}, \boldsymbol{n}') = \begin{cases} \lambda_d & \text{if } n_d < R_{od}, \boldsymbol{n}' = \boldsymbol{n} + \boldsymbol{e}_1, \\ \lambda_{hd} & \text{if } n_d \ge R_{od}, \boldsymbol{n}' = \boldsymbol{n} + \boldsymbol{e}_1, \\ \lambda_v & \text{if } n_v < N_v, \boldsymbol{n}' = \boldsymbol{n} + \boldsymbol{e}_2, \\ \lambda_{hv} & \text{if } N_v \le n_v < N_v + R_{hv}, \boldsymbol{n}' = \boldsymbol{n} + \boldsymbol{e}_2, \\ n_d\mu_d & \text{if } \boldsymbol{n}' = \boldsymbol{n} - \boldsymbol{e}_1, \\ n_v\mu_v & \text{if } \boldsymbol{n}' = \boldsymbol{n} - \boldsymbol{e}_2, \\ 0 & \text{in other cases,} \end{cases} \tag{2.49}$$

where $\lambda_v = \lambda_{ov} + \lambda_{hv}$.

Using the scheme of the proof of the Proposition 2.4, it is possible to show that the following fact is true (see Fig. 2.5, area in a dashed line).

Proposition 2.5 Stationary distribution of the system at use VP scheme has the following multiplicative form:

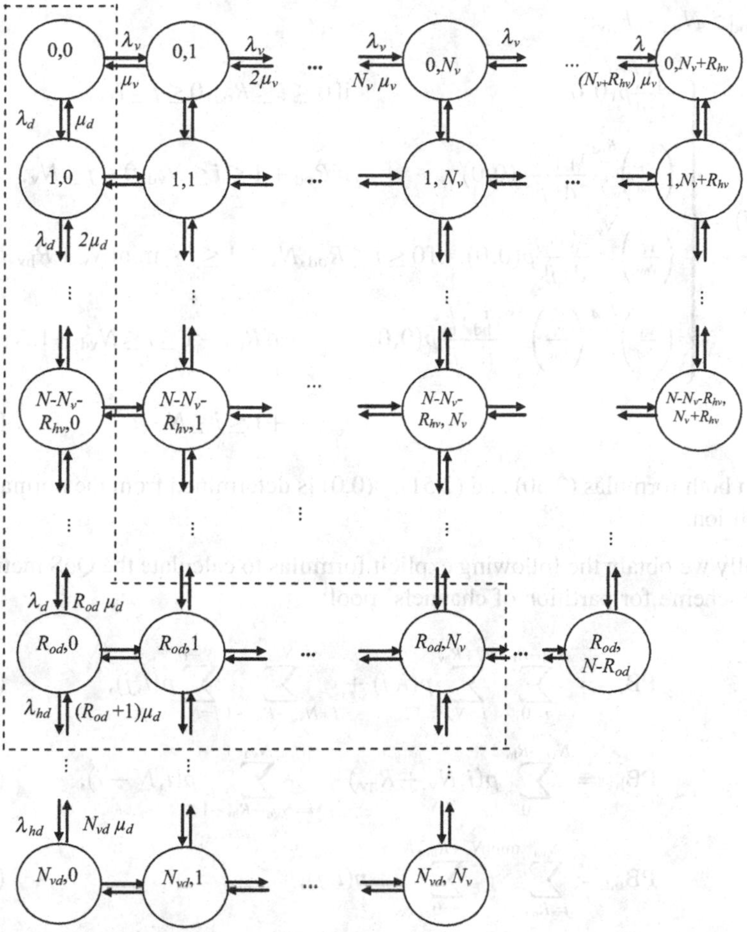

Fig. 2.5 State transition diagram of the model with VP scheme for partition of channels

Case $R_{\text{od}} \leq N_{\text{vd}} - R_{\text{hv}}$:

$$p(i,j) = \begin{cases} \dfrac{\nu_{\text{d}}^i}{i!}\dfrac{\nu_{\text{v}}^j}{j!}p(0,0) & \text{if } 0 \leq i \leq R_{\text{od}}, 0 \leq j \leq N_{\text{v}}, \\ \left(\dfrac{\nu_{\text{d}}}{\nu_{\text{hd}}}\right)^{R_{\text{od}}}\dfrac{\nu_{\text{hd}}^i}{i!}\dfrac{\nu_{\text{v}}^j}{j!}p(0,0) & \text{if } R_{\text{od}}+1 \leq i \leq N_{\text{vd}}, 0 \leq j \leq N_{\text{v}}, \\ \left(\dfrac{\nu_{\text{v}}}{\nu_{\text{hv}}}\right)^{N_{\text{v}}}\dfrac{\nu_{\text{d}}^i}{i!}\dfrac{\nu_{\text{hv}}^j}{j!}p(0,0) & \text{if } 0 \leq i \leq R_{\text{od}}, N_{\text{v}}+1 \leq j \leq N_{\text{v}}+R_{\text{hv}}, \\ \left(\dfrac{\nu_{\text{d}}}{\nu_{\text{hd}}}\right)^{R_{\text{od}}}\left(\dfrac{\nu_{\text{v}}}{\nu_{\text{hv}}}\right)^{N_{\text{v}}}\dfrac{\nu_{\text{hd}}^i}{i!}\dfrac{\nu_{\text{hv}}^j}{j!}p(0,0) & \text{if } R_{\text{od}}+1 \leq i \leq N_{\text{vd}}-1, N_{\text{v}} \\ & +1 \leq j \leq \min(N_{\text{v}}+R_{\text{hv}}, N-i); \end{cases} \quad (2.50)$$

Case $R_{od} > N_{vd} - R_{hv}$:

$$p(i,j) = \begin{cases} \frac{\nu_d^i}{i!}\frac{\nu_v^j}{j!}p(0,0) & \text{if } 0 \le i \le R_{od}, 0 \le j \le N_v, \\ \left(\frac{\nu_d}{\nu_{hd}}\right)^{R_{od}} \frac{\nu_{hd}^i}{i!}\frac{\nu_v^j}{j!}p(0,0) & \text{if } R_{od}+1 \le i \le N_{vd}, 0 \le j \le N_v, \\ \left(\frac{\nu_v}{\nu_{hv}}\right)^{N_v} \frac{\nu_d^i}{i!}\frac{\nu_{hv}^j}{j!}p(0,0) & \text{if } 0 \le i \le R_{od}, N_v+1 \le j \le \min(N_v+R_{hv}, N-i), \\ \left(\frac{\nu_d}{\nu_{hd}}\right)^{R_{od}} \left(\frac{\nu_v}{\nu_{hv}}\right)^{N_v} \frac{\nu_{hd}^i}{i!}\frac{\nu_{hv}^j}{j!}p(0,0) & \text{if } R_{od}+1 \le i \le N_{vd}-1, N_v+1 \le j \le N-i. \end{cases} \quad (2.51)$$

In both formulas (2.50) and (2.51), $p(0,0)$ is determined from the normalizing condition.

Finally we obtain the following explicit formulas to calculate the QoS metrics at use VP scheme for partition of channels' pool:

$$PB_{ov} = \sum_{i=0}^{N_{vd}-R_{hv}} \sum_{j=N_v}^{N_v+R_{hv}} p(i,j) + \sum_{i=N_{vd}-R_{hv}+1}^{N_{vd}} \sum_{j=N_v}^{N-i} p(i,j), \quad (2.52)$$

$$PB_{hv} = \sum_{i=0}^{N_{vd}-R_{hv}} p(i, N_v + R_{hv}) + \sum_{i=N_{vd}-R_{hv}+1}^{N_{vd}} p(i, N-i), \quad (2.53)$$

$$PB_{od} = \sum_{i=R_{od}}^{N_{vd}} \sum_{j=0}^{\min(N_v+R_{hv}, N-i)} p(i,j), \quad (2.54)$$

$$PB_{hd} = \sum_{i=0}^{N_v-1} p(N_{vd}, i) + \sum_{i=N_{vd}-R_{hv}}^{N_{vd}} p(i, N-i). \quad (2.55)$$

2.2.3 Numerical Results

The developed above explicit formulas allow to investigate the behavior of QoS metrics of the both partition schemes over any range of change of values of loading parameters of heterogeneous calls and number of channels. First of all, here it is assumed that allocation of entire pool of channels between zones is fixed, and only regulated parameters are R_{hv} and R_{od}. It is clear that the behavior of QoS metrics with respect to the indicated parameters is identical in both partition schemes. In other words, the increase in value of one of the parameters R_{hv} and R_{od} (in an admissible area) favorably influences the QoS metric of calls of the corresponding type only.

The initial data for total number of channels and loading parameters of heterogeneous calls are as in [4], i.e., $N = 30$, $\lambda_{ov} + \lambda_{hv} = 0.15$ call/s, $\lambda_{od} + \lambda_{hd} = 0.3$ call/s, $\mu_v^{-1} = 2$ s, and $\mu_d^{-1} = 120$ s. Below, assume that $N_v = 12$, $N_{vd} = 18$ and 30 % of the total intensity of voice calls are handover voice calls and 80 % of the total intensity of data calls are new data calls.

First consider the results of numerical experiments for the model with CP scheme for partition of channels. Some results for behavior of QoS metrics versus R_{hv} are shown in Fig. 2.6. Since loss probability of ov-calls is determined by Erlang's B-formula (i.e., it is independent on R_{hv}), then function PB_{ov} is constant

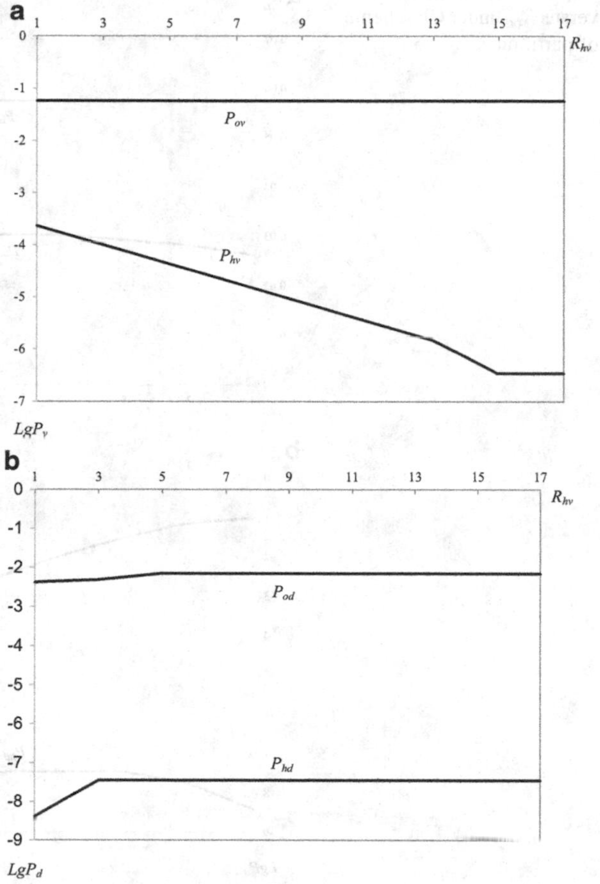

Fig. 2.6 QoS metrics versus R_{hv} under CP scheme of partition

one, while loss probability of hv-calls (PB_{hv}) is decreasing function versus R_{hv} (see Fig. 2.6a). Note that function PB_{hv} is almost piecewise linear one and has high rate of decreasing especially for small values of R_{hv}. Both QoS metrics for data calls are non-decreasing function with respect to R_{hv}, and they become almost constant for the large values of indicated parameter (see Fig. 2.6b). The last facts are explained by the following arguments: for given initial data, intensity of handover voice calls essentially is less than intensity of data calls, and at the same time handle rate of data calls essentially is more than appropriate parameter for voice calls.

Results for behavior of QoS metrics versus R_{od} are shown in Fig. 2.7. As above, function PB_{ov} is a constant one, but here PB_{hv} is a non-decreasing function, since increase in value of parameter R_{od} leads to decreasing the chances of hv-calls for access to the channels of vd-zone (see Fig. 2.7a). At that rate of change of function, PB_{hv} is inconsiderable for large values of parameter R_{od}. In this case, function PB_{od} is decreased with high speed in small values of parameter R_{od}, while function PB_{hd} has a small increasing rate in large values of indicated parameter.

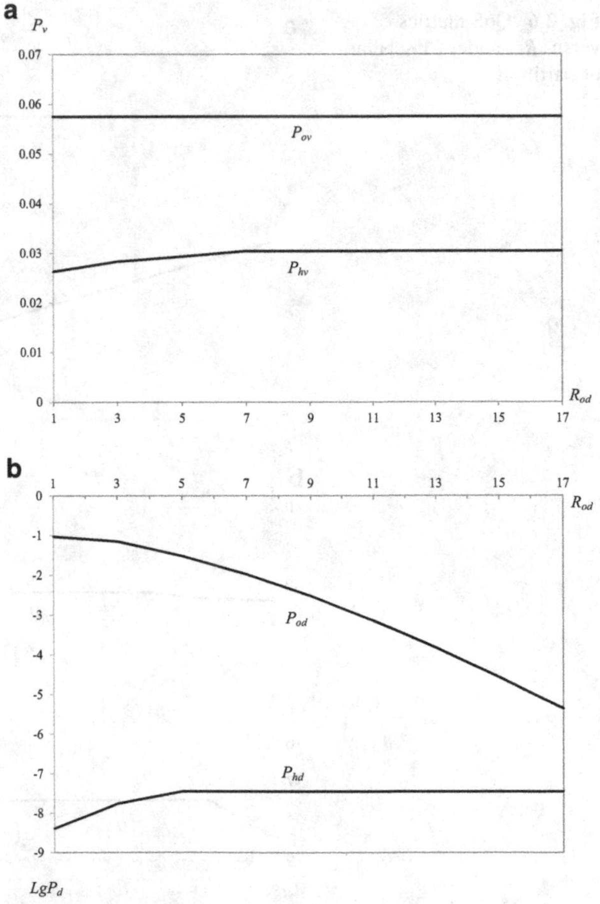

Fig. 2.7 QoS metrics versus R_{od} under CP scheme of partition

Now consider the results of numerical experiments for the model with VP scheme for partition of channels (see Figs. 2.8 and 2.9). In Fig. 2.8, the dependency of QoS metrics on the parameter R_{hv} is shown. It is seen from Fig. 2.8a that function PB_{hv} decreases in small values of parameter R_{hv} with high speed; thereafter, it becomes almost constant; function PB_{ov} increases with insignificant speed in small values of indicated parameter; thereafter, it becomes almost constant also. Almost constants are both functions PB_{od} and PB_{hd} versus R_{hv} (see Fig. 2.8b). Such behavior of functions PB_{od} and PB_{hd} is explained via small intensity of handover voice calls.

Dependency of QoS metrics on the parameter R_{od} is shown in Fig. 2.9. Here both functions PB_{od} and PB_{hv} increase with insignificant speed in small values of indicated parameter; thereafter, it becomes almost constant (see Fig. 2.9a). However, function PB_{od} decreases with significant speed versus PB_{od} and PB_{od}, while function PB_{od} and PB_{hd} is almost constant one (see Fig. 2.9b).

Fig. 2.8 QoS metrics versus R_{od} under VP scheme of partition

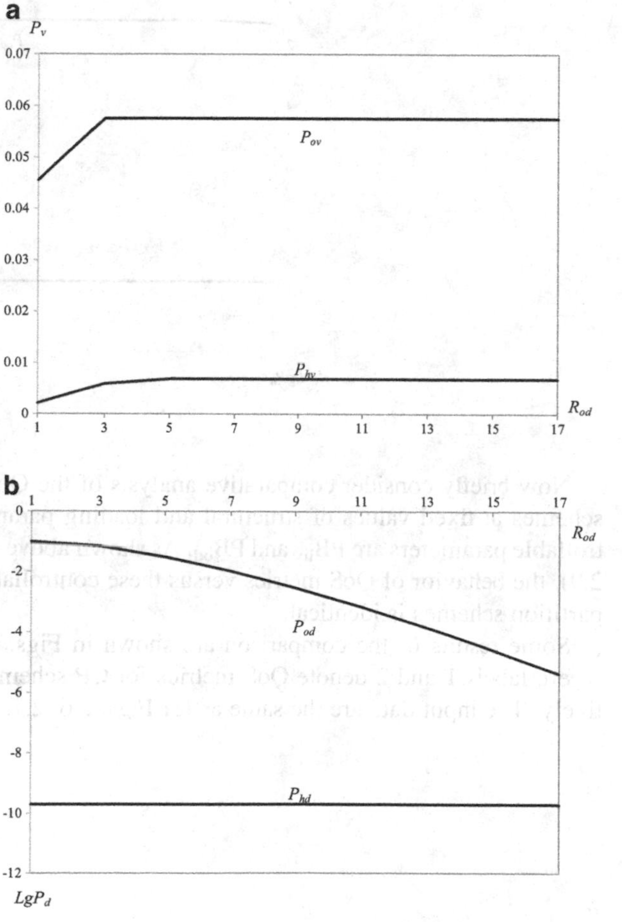

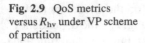

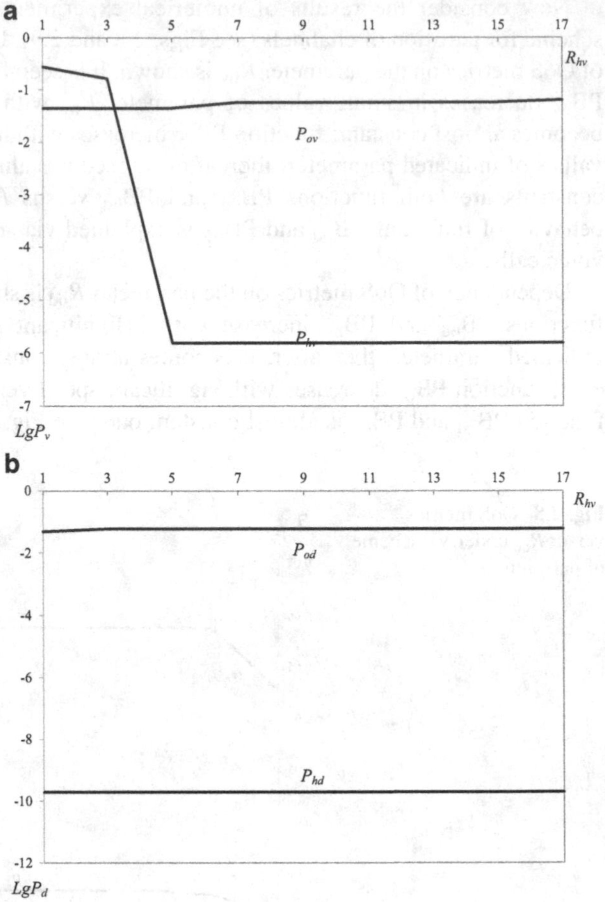

Fig. 2.9 QoS metrics versus R_{hv} under VP scheme of partition

Now briefly consider comparative analysis of the QoS metrics of two partition schemes at fixed values of structural and loading parameters of the model. Controllable parameters are PB_{hv} and PB_{od}. As shown above (see Figs. 2.6, 2.7, 2.8, and 2.9), the behavior of QoS metrics versus these controllable parameters in different partition schemes is identical.

Some results of the comparison are shown in Figs. 2.10, 2.11, 2.12, and 2.13 where labels 1 and 2 denote QoS metrics for CP scheme and VP scheme, respectively. The input data are the same as for Figs. 2.6, 2.7, 2.8, and 2.9.

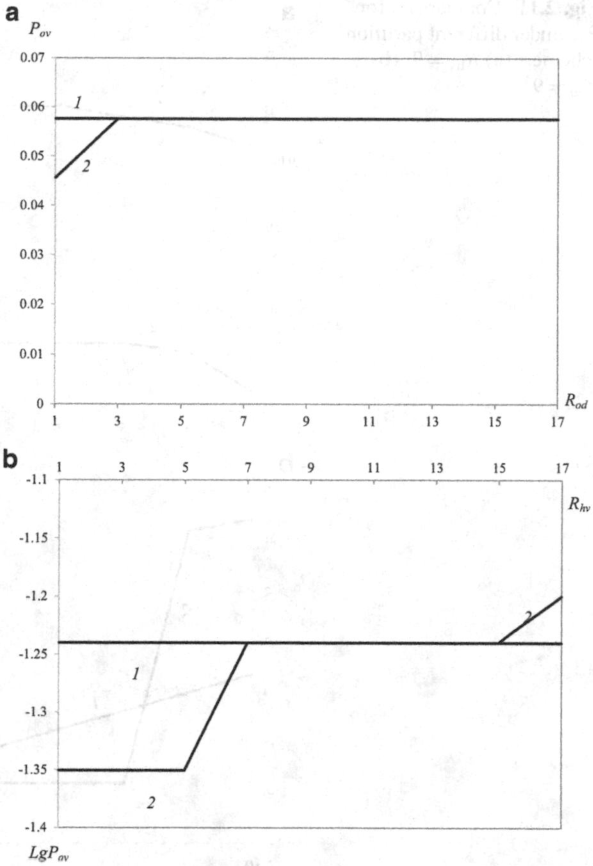

Fig. 2.10 Comparison for P_{ov} under different partition schemes; (**a**) $R_{hv} = 9$; (**b**) $R_{od} = 9$

It is interesting to note that character of change of QoS metrics versus change of parameters PB_{hv} and PB_{od} is almost identical at both partition schemes of channels (except for the QoS metric PB_{hv} versus R_{hv}, see Fig. 2.11b). However, in some cases, their absolute values are in different quantitative ranges.

From Fig. 2.10a, we conclude that for the chosen initial data, QoS metric PB_{ov} is better under VP scheme of partition for values $R_{od} \leq 2$ and for $R_{od} \geq 3$ both partition schemes have the same performance. However, from Fig. 2.10b it is seen that this QoS metric is better under VP scheme of partition for values $R_{od} \leq 8$ and in cases $R_{od} > 9$ favorably scheme for QoS metric PB_{ov} is CP scheme of partition.

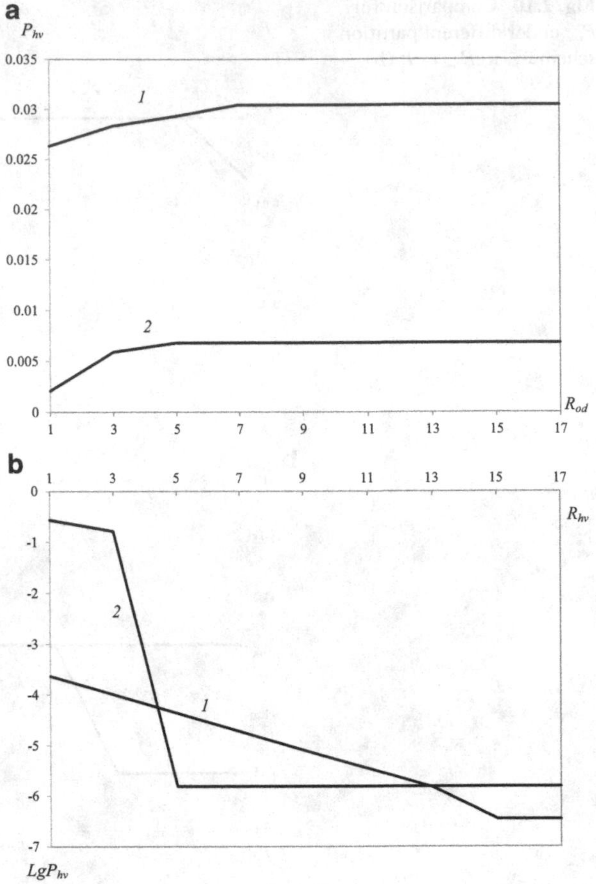

Fig. 2.11 Comparison for P_{hv} under different partition schemes; (**a**) $R_{hv}=9$; (**b**) $R_{od}=9$

In Fig. 2.11 comparative results are shown for QoS metric PB_{hv}. It is seen from Fig. 2.11a that this QoS metric is essentially better under VP scheme of partition for all values of R_{od}. However, this QoS metric is better under VP scheme for values of $R_{hv} \in [3, 7]$, and in other values this metric favorably is CP scheme (see Fig. 2.12b).

It is seen from Fig. 2.12a that for QoS metric PB_{od} at $R_{od} < 12$ both partition schemes have the same performance, but at $R_{od} \geq 12$ one of schemes, i.e., VP scheme, has good performance for this metric. Note that this QoS metric is essentially better under CP scheme of partition for all values of R_{hv} (see Fig. 2.12b).

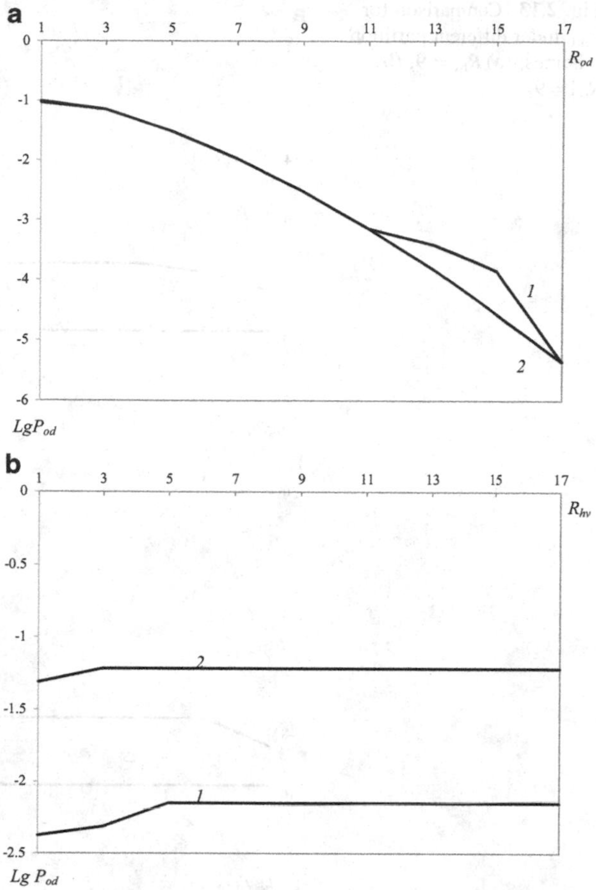

Fig. 2.12 Comparison for P_{od} under different partition schemes; (**a**) $R_{hv} = 9$; (**b**) $R_{od} = 9$

From Fig. 2.13 we conclude that QoS metric PB_{hd} is essentially better under VP scheme of partition for all values of both parameters R_{od} and R_{hv}.

The numerical results show that all QoS metrics in both partition schemes have monotony property. These facts allow to develop the algorithms to finding the set of effective values (SEVs) in order to satisfy the given QoS level. Such kind of problems has been considered in Sect. 2.1.1; thus they are not considered here. We left them to reader.

Fig. 2.13 Comparison for P_{hd} under different partition schemes; **(a)** $R_{hv}=9$; **(b)** $R_{od}=9$

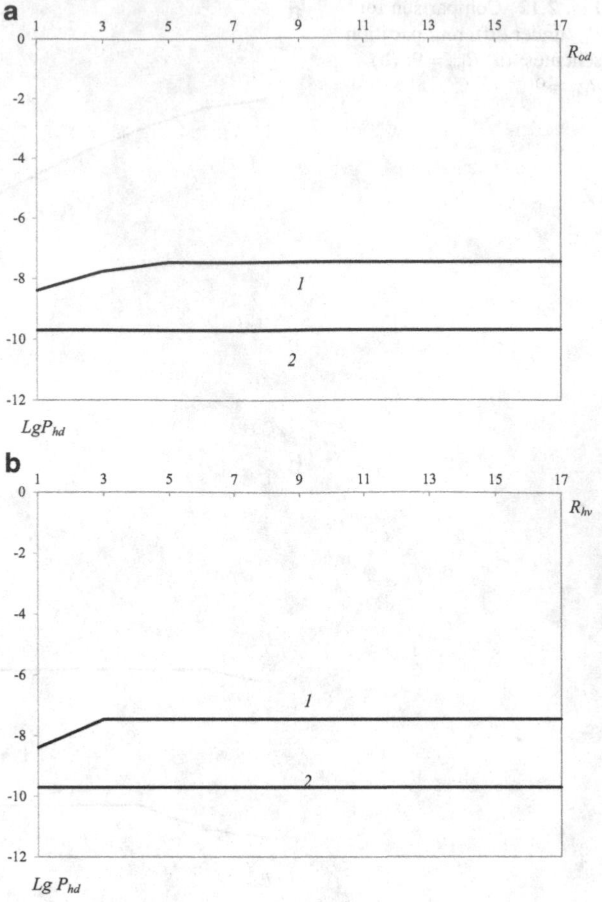

2.3 Conclusion

In the last decades are published many books [1, 2, 5, 6, 10–12, 19, 27, 30, 31, 33] and reviews [3, 14, 15, 24] which deal with applications of queueing theory in telecommunication networks. In this direction, multi-rate Erlang's models are subjects of many researches. Rather detailed description of the MRQ results can be found in appropriate chapters of books [11, 19, 27, 30] and in papers [14, 15, 23, 26]. Here we only note that in the theory of MRQ models, the main results are based on the following fact [13, 29]: the stationary distribution of unbuffered Markov model of MRQ in the full availability of channels (i.e., in CS scheme) has a multiplicative form. Note that as in case of single-rate multidimensional Erlang's models, a multiplicative solution exists also in the cases when the distribution function of the service time of multi-rate traffic is arbitrary with the fixed mean value.

In this chapter an analytical approach to the analysis of a multi-rate Erlang's model with the state-dependent randomized access strategy is proposed [21, 23]. In contrast to the known numerical methods, the proposed approach does not require generation of the large state space of the model, and, therefore, finding the desired QoS metrics is carried out by explicit formulas. It is shown that from the results of this chapter, the results for the MRQ models with the known access schemes can be easily obtained. It is proved that the results, obtained in previous studies using heuristic considerations, which have been regarded as approximate, are in fact accurate ones in some cases. The proposed method also allows one to develop simple one-dimensional recurrence formulas for calculating QoS metrics of integral wireless networks of voice calls and data calls. The simplicity of the obtained formulas allows one to formulate and solve important problems of optimization of the studied models.

Note that Kaufman–Roberts algorithm together with its various modifications is the main tool to study the characteristics of multi-rate queues. For instance, in [8] the formula to calculate the occupancy distribution in the CS scheme for the model MRQ with mixture of flows of Erlang, Engset, and Pascal type is proposed. Here the considered models of MRQ systems with randomized access strategy have been studied in [31], Chap. 7, where they are called state-dependent systems. In the indicated book, the equations (2.18) are obtained also. However, in [31] these equations are obtained subject to the following conditions: $(a_i(n))/(a_i(n+b_j)) = (a_j(n))/(a_j(n+b_i))$ for any $i, j = 1, 2, \ldots, K$. The last conditions are necessary to providing the reversibility of appropriate K-dimensional Markov chain which are results from Kolmogorov's theorem [18]. At the same time, as authors noted, in practice the fulfillment of these conditions is extremely difficult problem. The equations (2.18) without any proofs are resulted in the book [19], Chap. 11 also. Authors simply verbally assume that these equations are true. Alternative and effective approach to calculate the QoS metrics of multi-rate models is using convolution algorithms [11].

Attempt to the solution of similar problems which appear in communication networks has been made in work [32]. Unfortunately, this attempt has appeared unsuccessful [22].

Here two partition schemes for distribution of entire pool of channels among voice and data calls in integral wireless networks are proposed also. One of them uses isolated (rigid) distribution of channels, while in another scheme, virtual distribution procedure is applied. In both schemes, a voice call seizes the free channel in own zone, and if there is no free channel in this zone, only handover voice calls might search free channel in another zone. Moreover the state-dependent limit to the both number of handover voice calls and new data calls in zone of channels for data calls are defined. It is shown that in both partition schemes stationary distribution of appropriate 2-D MC has multiplicative form. By using this fact the explicit formulas to calculate the QoS metrics of the integral networks under given partition schemes are developed. The proposed formulas allow to perform comparative analysis of QoS metrics in various partition schemes.

References

1. Akimura H, Kawashima M (2003) Teletraffic theory and applications. Springer, London
2. Andreadis A, Giambene G (2003) Protocols for high efficiency wireless networks. Kluwer, Boston
3. Basharin GP, Samouylov KE, Yarkina NV, Gudkova IA (2009) A new stage in mathematical teletraffic theory. Autom Rem Contr 70(12):1954–1964
4. Carvalho GHS, Martins VS, Frances CRL, Costa JCWA, Carvalho SV (2008) Performance analysis of multi-service wireless network: an approach integrating CAC, scheduling, and buffer management. Comput Electr Eng 34:346–356
5. Chen H, Huang L, Kumar S, Kuo CC (2004) Radio resource management for multimedia QoS supports in wireless networks. Kluwer, Boston
6. Daigle JN (2005) Queueing theory with applications to packet telecommunication. Springer, New York
7. Delaire M, Hebuterne G (1997) Call blocking in multi-services systems on one transmission link. In: Proceedings of 5th international workshop on performance model and evaluation of ATM networks, July, UK, pp 253–270
8. Delbrouck LEN (1983) On the steady-state distribution in a service facility carrying mixtures of traffic with different peakedness factors and capacity requirements. IEEE Trans Commun 31(11):1209–1211
9. Feng W, Kowada M (2008) Performance analysis of wireless mobile networks with queueing priority and guard channels. Int Trans Oper Res 15:481–508
10. Giambene G (2005) Queueing theory and telecommunications. Networks and applications. Springer, New York
11. Iversen VB (2010) Teletraffic engineering and network planning. Technical University of Denmark, Lyndby
12. Janevski T (2003) Traffic analysis and design of wireless IP networks. Artech House, Boston
13. Kaufman JS (1981) Blocking in shared resource environment. IEEE Trans Commun Technol 10(10):1474–1481
14. Kelly FP (1991) Loss networks. Ann Appl Probab 1(3):319–378
15. Kelly FP (1986) Blocking probabilities in large circuit-switched networks. Adv Appl Probab 18:477–505
16. Kim CS, Melikov AZ, Ponomarenko LA (2009) Numerical investigation of a multi-threshold access strategy in multi-service cellular wireless networks. Cybern Syst Anal 45(5):680–691
17. Kim CS, Melikov AZ, Ponomarenko LA, Baek JH (2011) An analytical approach for performance analysis of multi-service cellular wireless networks with a randomized access strategy. Appl Comput Math 10(3):529–538
18. Kolmogorov A (1936) Zum theorie der Markoffschen ketten. Mathematische Annalen B112:155–160
19. Lakatos L, Szeidl L, Telek M (2013) Introduction to queueing systems with telecommunication applications. Springer, London
20. Melikov AZ, Fattakhova MI, Babayev AT (2005) Investigation of cellular communication networks with private channels for service of handover calls. Automat Contr Comput Sci 39 (3):61–69
21. Melikov AZ, Ponomarenko LA (2011) Multidimensional Erlang model with randomized call admission control strategy and its application in communication networks. Cybern Syst Anal 47(4):606–612
22. Melikov AZ, Ponomarenko LA (2013) Comment on "Probabilistic framework and performance evaluation for prioritized call admission control in next generation networks". Comput Commun 36(3):360–361
23. Melikov AZ, Ponomarenko LA, Kim CS (2011) Multirate Erlang's model with randomized access strategy and its application in multiservice communication networks. J Autom Inform Sci 43(2):23–38

24. Neimann VI (2009) Teletraffic and queueing theory. Autom Rem Contr 70(12):1965–1973
25. Ogbonmwan SE, Wei L (2006) Multi-threshold bandwidth reservation scheme of integrated voice/data wireless networks. Comput Commun 29(9):1504–1515
26. Oh Y, Kim CS, Melikov AZ, Fattakhova MI (2010) Numerical analysis of multi-parameter strategy of access in multi-service cellular communication networks. Autom Rem Contr 71 (12):2558–2572
27. Ponomarenko L, Kim CS, Melikov A (2010) Performance analysis and optimization of multi-traffic on communication networks. Springer, Heidelberg
28. Pioro M, Lubacs J, Korner U (1990) Traffic engineering problems in multi-service circuit-switched networks. Comput Netw ISDN Syst 20(1–5):127–136
29. Roberts JW (1981) A service system with heterogeneous user requirements application to multi-service telecommunication systems. In: Pujolle G (ed) Performance of data communication systems and their application. North Holland, Amsterdam, pp 423–431
30. Ross KW (1995) Multi-service loss models for broadband telecommunication networks. Springer, New York
31. Stasiak M, Glabowski M, Wishniewski A, Zwierzykowski P (2011) Modeling and dimensioning of mobile networks. From GSM to LTE. Wiley, Chichester
32. Tsiropoulos GI, Stratogiannis DG, Kanellopoulos SD, Cottis PG (2011) Probabilistic framework and performance evaluation for prioritized call admission control in next generation networks. Comput Commun 34:1045–1054
33. Yue W, Matsumoto Y (2002) Performance analysis of multi-channel and multi-traffic on wireless communication networks. Kluwer, Boston

Chapter 3
Algorithmic Methods for Analysis of Mono-service Cellular Networks

In this chapter, methods for exact and approximate evaluation of the characteristics of mono-service cellular networks with single traffic are developed. In other words, models with only one type of traffic in which new and handover calls have the same bandwidth requirements and identically distributed channel holding time with common average value are investigated. Two kinds of models have been considered. In models with queues (finite or infinite) of both new and handover calls, it is assumed that unlike calls may leave the queue if their waiting time is greater than some threshold value. Another type of models takes into account repetitions of new calls, while to provide high priority to handover calls, guard channel scheme is used. For both types of models results of numerical experiments are presented.

3.1 Models of Mono-service Cellular Networks with Buffers

To support the given level of QoS cellular networks employ various strategies for the access to radio channels of base stations (BSs) and/or organize buffer stores to wait for unlike calls either in queue or in an orbit (retrial queues).

Since h-calls are more sensitive to possible losses and delays than o-calls, the proposed schemes often imply that guard channels are used for h-calls and/or only their queues in the base station are organized. A queue of h-calls can be organized in networks where microcells are covered with some macrocell, i.e., there is some zone (handover zone) inside which a mobile user (MU) can be served in any of the neighboring cells. The time it takes an MU to cross the handover zone is called degradation interval. When an MU arrives at a handover zone, the presence of free channels in a new cell is checked. If there is a free channel, the h-call immediately occupies it, and the handover procedure is considered to be successfully completed at this stage; otherwise the channel of the previous cell is still used by this h-call and

A. Melikov, L. Ponomarenko, *Multidimensional Queueing Models in Telecommunication Networks*, DOI 10.1007/978-3-319-08669-9_3

is simultaneously queued to wait until any channel of the new cell becomes free. If a free channel does not appear before the end of the degradation interval, the conversation of the h-call is forcedly interrupted.

Note that to compensate for chances of o-calls in some networks, buffers are organized for this type of calls as well. Alternative way is using access scheme with retrial new calls. The last scheme is characterized by the feature that arriving calls who find all channels busy join the retrial group (orbit) to try again for their requests in random order and at random intervals. Obviously, both schemes increase the total throughput of the network.

In this section we study models with buffers for both types of calls as well as models with retrial o-calls where h-calls are handled in accordance with guard channel scheme.

Models with finite buffers are analyzed in the literature. However, the known approaches allow analyzing models with only a small buffer store. In this connection, here we propose an approach to analyzing both models with arbitrary volume of finite buffer store and models with infinite buffer store. One more advantage of the proposed approach is that, in contrast to the known approaches, it allows deriving simple analytic formulas to evaluate the unknown QoS metrics of networks under study. Note that though we consider here mono-service networks (to simplify model description and intermediate computations), the results obtained can easily be adapted to multiservice networks.

3.1.1 Models with Buffers

Here, consider a model of an isolated cell of a wireless network whose base station contains $N > 1$ radio channels. We assume that o-calls (h-calls) arrive according to the Poisson law with the rate $\lambda_o(\lambda_h)$, and the time a channel is occupied with calls of any type is an exponentially distributed random variable with the mean value μ^{-1}. If a handover takes place during servicing of a call of any type, the residual service time of a call in a new cell is also exponentially distributed with the same mean value since exponential distribution has no memory.

First consider model with finite separate buffers for unlike calls. These calls are served with channel reservation scheme for h-calls, i.e., an o-call arrived is accepted only if the number of free radio channels of the BS is greater than g, $0 \leq g \leq N-1$ (or, equivalently, if the number of busy radio channels of the BS is less than $N-g$). Otherwise the o-call is queued if the number of such calls in the corresponding buffer does not exceed a prescribed value R_o, where $0 < R_o < \infty$; otherwise the o-call arrived is blocked. A handover call is accepted if there is at least one free channel. Otherwise the h-call is queued if the number of such calls in the corresponding buffer does not exceed R_h, where $0 < R_h < \infty$; otherwise the h-call arrived is blocked.

Channel allocation scheme for a call to be chosen from the queue is defined as follows. If the number of free channels in the BS at this moment is g, then one o-call is chosen from the queue (if any) for service; otherwise, the free channel stands idle even if a queue of o-calls exists. Channels standing idle are inadmissible if there are h-calls in the cell. Any service discipline can be used inside each queue; for the sake of determinacy, we imply the first-come first-served (FCFS) discipline.

Consider models with impatient o-calls. This means that an o-call may leave the queue before the service starts if its time of waiting in the buffer exceeds a random variable with finite mean value τ_0^{-1}. Similarly, an h-call may leave the queue before the service starts if the time of its degradation (i.e., the time it takes to cross the handover zone) exceeds a random variable with the finite mean value τ_h^{-1}. The above random variables are assumed to be independent of each other and equally exponentially distributed.

The main metrics of this model are loss probabilities for unlike calls, their average queue lengths, and the mean waiting time for unlike calls. Below, both the exact and approximate methods to calculate QoS metrics of models under study are developed.

3.1.2 Models with Finite Buffers

Let us first consider a model with finite queues of unlike calls. To describe the operation of the system under study, a two-dimensional Markov chain (MC) is used. The state of a cell at an arbitrary time can be defined by a two-dimensional vector $\boldsymbol{k} = (k_1, k_2)$, where k_1 indicates the total number of busy channels and h-calls in the queue and k_2 is the number of o-calls in the queue. The state space of the system is defined as follows:

$$S = \cup_{i=0}^{R_o} S_i, \tag{3.1}$$

where

$$S_0 = \{\boldsymbol{k} : k_1 = 0, 1, \ldots, N + R_h;\ k_2 = 0\},$$

$$S_i = \{\boldsymbol{k} : k_1 = N - g, \ldots, N + R_h;\ k_2 = i\},\ i \geq 1.$$

Considering the mechanism of system operation, we find that the nonnegative elements of the given 2-D MC are determined as follows (see Fig. 3.1):

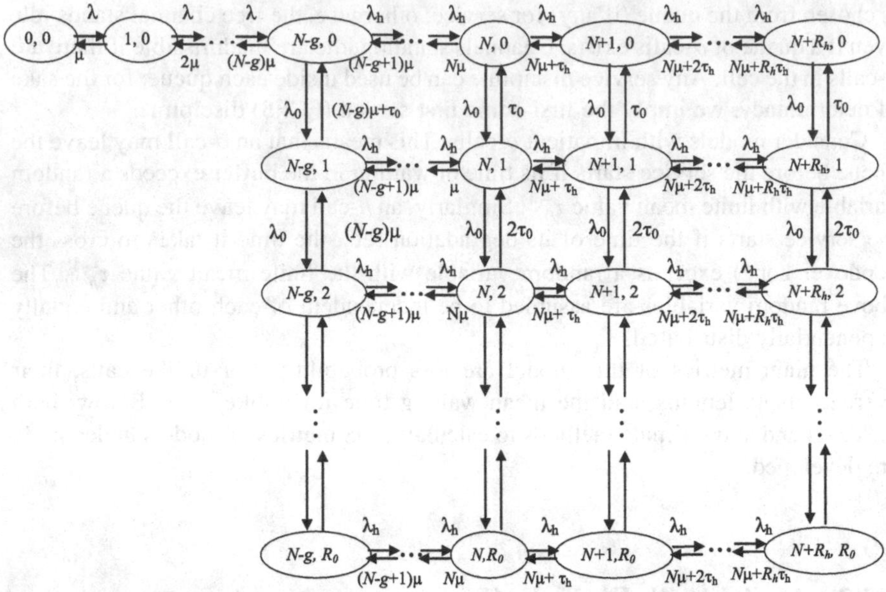

Fig. 3.1 State diagram of the model with guard channels and finite buffers for impatient calls

$$q(\boldsymbol{k},\boldsymbol{k}') = \begin{cases} \lambda_o + \lambda_h & \text{if } k_1 < N-g,\ k_2 = 0,\ \boldsymbol{k}' = \boldsymbol{k} + \boldsymbol{e}_1, \\ \lambda_o & \text{if } k_1 \geq N-g,\ \boldsymbol{k}' = \boldsymbol{k} + \boldsymbol{e}_2, \\ \lambda_h & \text{if } k_1 \geq N-g,\ \boldsymbol{k}' = \boldsymbol{k} + \boldsymbol{e}_1, \\ f(k_1)\mu + (k_1 - N)^+\tau_h & \text{if } \boldsymbol{k}' = \boldsymbol{k} - \boldsymbol{e}_1, \\ (N-g)\mu\delta(k_1, N-g) + k_2\tau_o & \text{if } \boldsymbol{k}' = \boldsymbol{k} - \boldsymbol{e}_2 \\ 0 & \text{in other cases.} \end{cases} \tag{3.2}$$

Hereinafter, we introduce the following notation: $f(x) = \min(x, N)$, $x^+ = \max(0, x)$.

The unknown QoS metrics are determined via the stationary distributions of the state probabilities of the model. So, the average number of o-calls (L_o) and the average number of h-calls (L_h) in the queue are determined as the corresponding marginal distributions of the original chain:

$$L_o = \sum_{k_2=1}^{R_o} k_2 \sum_{k_1=N-g}^{N+R_h} p(k_1, k_2), \tag{3.3}$$

$$L_h = \sum_{k_1=N+1}^{N+R_h} (k_1 - N) \sum_{k_2=0}^{R_o} p(k_1, k_2). \tag{3.4}$$

To find the loss probability for unlike calls, the following approach can be used. As indicated above, o-calls are lost in case of the following events: (1) at the time an o-call arrives, there are already R_o such calls in the corresponding buffer; (2) the time an o-call is waiting in the buffer exceeds a prescribed threshold τ_o^{-1}. Hence, the loss probability for o-calls (P_o) can be determined as follows:

$$P_o = \sum_{k_1=N-g}^{N+R_h} p(k_1, R_o) + \frac{1}{\lambda_o} \sum_{k_2=1}^{R_o} k_2 \tau_o \sum_{k_2=N-g}^{N+R_h} p(k_1, k_2). \tag{3.5}$$

The first and second terms of the sum in the last formula denote the probability of events (1) and (2), respectively.

Similarly, we conclude that h-calls are lost in case of the following events: (3) at the time an h-call arrives, there are already R_h such calls in the corresponding buffer; (4) the degradation interval for an h-call ends earlier than it gets access to the free channel.

Hence, the loss probability for h-calls (P_h) can be determined as follows:

$$P_h = \sum_{k_2=0}^{R_o} p(N + R_k, k_2) + \frac{1}{\lambda_h} \sum_{k_1=N+1}^{N+R_h} (k_1 - N)\tau_h \sum_{k_2=0}^{R_o} p(k_1, k_2). \tag{3.6}$$

Formula (3.6) can be commented similarly to Eq. (3.5). Further, formulas (3.3)–(3.6) and a modified Little's formula can be used to evaluate the mean waiting time for o-calls (W_o) and h-calls (W_h) in the buffer:

$$W_x = \frac{L_x}{\lambda_x(1 - P_x)}, \; x \in \{o, h\}. \tag{3.7}$$

Thus, to find QoS metrics (3.3)–(3.7), it is necessary to determine the steady-state probabilities of the model from the corresponding SGBE. It is composed based on relations (3.2); its explicit form and a solution algorithm for the problem are presented in [7]. Such approach to calculating the QoS metrics is called exact one.

However, being combinatory, this approach is efficient only for small values of R_o and R_h and is absolutely unsuitable even for their moderate values. At the same time, from the practical standpoint, of interest are models with arbitrary size of buffer stores for waiting unlike calls (they are also of certain scientific interest).

In view of the above facts, to overcome the mentioned computational difficulties, below we propose to use an approximate method based on the principles of state space merging of 2-D MC.

This method is applicable to analyze models of popular microcells, with the intensity of h-calls being much greater than the intensity of o-calls. In other words, below we assume that $\lambda_h \gg \lambda_o$. It is important to note that this assumption is not extraordinary for cellular networks, since this is a regime that commonly occurs in microcells, in which mobile users have high mobility and short duration calls [6]. Indeed in 3.5G wireless network (e.g., IEEE 802.16e), the radius of a microcell is about 100 m, and so the arrival rate of h-calls is larger than that of o-calls, and holding time in a microcell is quite short. Thus microcell in 3.5G wireless networks satisfies the above conditions. Moreover, as is seen from the further presentation, final results do not depend directly on loading parameters of incoming traffic but only on their relations $v_x := \lambda_x/\mu, x \in \{o, h\}$.

By the above assumptions on the relationships between loading parameters of the traffic of different types, we find that in representation (3.1), the transition rates between states inside each class S_i much exceed the rates of transitions between classes. Based on this, the sets S_i are then united into individual merged states $\langle i \rangle$, and the following merging function with the domain (3.1) is introduced:

$$U(\boldsymbol{k}) = \langle i \rangle \text{ if } \boldsymbol{k} \in S_i,\ i = 0, 1, \ldots, R_o. \tag{3.8}$$

The merged function (3.8) defines a merged model, which is a 1-D MC with the finite state space $\widetilde{S} := \{\langle i \rangle : i = 0, 1, 2, \ldots, R_o\}$.

To find the stationary distribution of the original model, a preliminary determination of stationary distributions of split models is required. A split model with the state space S_o is described by a 1-D BDP whose parameters are defined as follows (see Eq. (3.2)):

$$\lambda_j = \begin{cases} \lambda_o + \lambda_h & \text{if } j < N - g, \\ \lambda_h & \text{if } j \geq N - g; \end{cases} \quad \mu_j = \begin{cases} j\mu & \text{if } j \leq N, \\ N\mu + (j - N)\tau_h & \text{if } j > N. \end{cases} \tag{3.9}$$

The probabilities of states in this split model are denoted by $\rho_o(i)$, $i = 0, 1, 2, \ldots, N + R_h$. Considering Eq. (3.9), they can be determined as follows:

$$\rho_0(i) = \begin{cases} \dfrac{v^i}{i!}\rho_0(0) & \text{if } 1 \leq i \leq N - g, \\ \left(\dfrac{v}{v_h}\right)^{N-g} \dfrac{v_h^i}{i!}\rho_0(0) & \text{if } N - g + 1 \leq i \leq N, \\ \dfrac{v^{N-g}}{N!} v_h^g \prod_{j=N+1}^{i} \dfrac{\lambda_h}{N\mu + (j - N)\tau_h}\rho_0(0) & \text{if } N + 1 \leq i \leq N + R_h, \end{cases} \tag{3.10}$$

where $v = v_o + v_h$ and $\rho_0(0)$ are defined from normalizing condition, i.e., $\sum_{i=0}^{N+R_h} \rho_0(i) = 1$.

Split models with the state space S_i are 1-D BDP, identical for all $i \geq 1$. The birth rate is a constant and equal to λ_h, and the death rate in the state j is equal to $f(j)\mu + (j - N)^+\tau_h$, where $j = N - g, \ldots, N + R_h$. Hence, steady-state probabilities of split models with the state space S_i, $i \geq 1$, denoted by $\rho_i(j)$, can be calculated as follows (since all the split models with the state space S_i, $i \geq 1$ have identical distributions, the index i below in the notation $\rho_i(j)$ is omitted):

$$\rho(j) = \begin{cases} \dfrac{v_h^j}{j!}\dfrac{(N-g)!}{v_h^{N-g}}\rho(N-g) & \text{if} N-g+1 \le j \le N, \\ v_h^g \dfrac{(N-g)!}{N!}\prod_{i=N+1}^{j}\dfrac{\lambda_h}{N\mu + (i-N)\tau_h}\rho(N-g) & \text{if} N+1 \le j \le N+R_h, \end{cases} \tag{3.11}$$

where $\rho(N-g)$ is defined from normalizing condition, i.e., $\sum_{i=N-g}^{N+R_h}\rho(i) = 1$.

To find the stationary distribution $\pi(\langle i\rangle)$, $\langle i\rangle \in \widetilde{S}$ of the merged model, it will suffice to determine its generating matrix. Considering Eqs. (3.2), (3.10), and (3.11), we find that the elements of indicated matrix $q(\langle i\rangle, \langle j\rangle)$, $\langle i\rangle$, $\langle j\rangle \in \widetilde{S}$ can be determined from the following relationships:

$$q(\langle i\rangle, \langle j\rangle) = \begin{cases} \lambda_o^* & \text{if } i=0,\ j=1, \\ \lambda_o & \text{if} i>0,\ j=i+1, \\ ((N-g)\mu + i\tau_o)\rho(N-g) + i\tau_o(1-\rho(N-g)) & \text{if} j=i-1, \\ 0 & \text{in other cases}, \end{cases} \tag{3.12}$$

where $\lambda_o^* = \lambda_o(1 - \sum_{i=0}^{N-g-1}\rho_0(i))$.

Hence, the state probabilities of the merged model can be determined as the stationary distribution of the 1-D BDP with the rates specified by Eq. (3.12), i.e.,

$$\pi(\langle j\rangle) = \frac{\lambda_o^*\lambda_o^{j-1}}{\prod_{i=1}^{j} q(\langle i\rangle, \langle i-1\rangle)}\pi(\langle 0\rangle), \quad j = 1, \ldots, R_o, \tag{3.13}$$

where $\pi(\langle i\rangle)$ is defined from normalizing condition, i.e., $\sum_{i=0}^{R_o}\pi(\langle i\rangle) = 1$.

Considering Eqs. (3.10)–(3.13), the stationary distribution of the original model can approximately be found as follows:

$$\begin{aligned} &p(0,k_2) \approx \rho_0(k_2)\pi(\langle 0\rangle); \\ &p(k_1,k_2) \approx \rho(k_2)\pi(\langle k_1\rangle), \ k_1 \ge 1. \end{aligned} \tag{3.14}$$

Then with Eqs. (3.3) and (3.14) we find that the average number of o-calls in the queue is defined by

$$L_o \approx \sum_{i=1}^{R_o} i \sum_{j=N-g}^{N+R_h}\rho(j)\pi(\langle i\rangle) = \sum_{i=1}^{R_o} i\pi(\langle i\rangle)\sum_{j=N-g}^{N+R_h}\rho(j) = \sum_{i=1}^{R_o} i\pi(\langle i\rangle). \tag{3.15}$$

Let us write the average number of h calls in the queue as follows (see Eq. (3.4)):

$$
\begin{aligned}
L_h \approx \sum_{i=1}^{R_h} i \sum_{j=0}^{R_o} \rho_j(N+i)\pi(\langle j\rangle) &= \sum_{i=1}^{R_h} i\left(\rho_0(N+i)\pi(\langle 0\rangle) + \sum_{j=1}^{R_o}\rho(N+i)\pi(\langle j\rangle)\right)\\
&= \sum_{i=1}^{R_h} i\left(\rho_0(N+i)\pi(\langle 0\rangle) + \rho(N+i)\sum_{j=1}^{R_o}\pi(\langle j\rangle)\right)\\
&= \sum_{i=1}^{R_h} i(\rho_0(N+i)\pi(\langle 0\rangle) + \rho(N+i)(1-\pi(\langle 0\rangle))).
\end{aligned}
\tag{3.16}
$$

The loss probability for o-calls can approximately be determined as follows (see Eq. (3.5)):

$$
\begin{aligned}
P_o &\approx \sum_{i=N-g}^{N+R_h} \rho_{R_h}(i)\pi(\langle R_o\rangle) + \frac{\tau_o}{\lambda_o}\sum_{j=1}^{R_o} j \sum_{i=N-g}^{N+R_h} \rho(i)\pi(\langle j\rangle)\\
&= \pi(\langle R_o\rangle) + \frac{\tau_o}{\lambda_o}\sum_{j=1}^{R_o} j\pi(\langle j\rangle).
\end{aligned}
\tag{3.17}
$$

Similarly, we find the following approximate formula to calculate the loss probability for h-calls (see Eq. (3.6)):

$$
\begin{aligned}
P_h &\approx \sum_{i=0}^{R_o} \rho_i(N+R_h)\pi(\langle i\rangle) + \frac{\tau_h}{\lambda_h}\sum_{i=N+1}^{N+R_h}(i-N)\sum_{j=0}^{R_o}\rho_j(i)\pi(\langle j\rangle)\\
&= \rho_0(N+R_h)\pi(\langle 0\rangle) + \rho(N+R_h)\sum_{i=1}^{R_o}\pi(\langle i\rangle)\\
&\quad + \frac{\tau_h}{\lambda_h}\sum_{i=1}^{R_h} i\sum_{j=0}^{R_o}\rho_j(N+i)\pi(\langle j\rangle)\\
&= \rho_0(N+R_h)\pi(\langle 0\rangle) + \rho(N+R_h)(1-\pi(\langle 0\rangle))\\
&\quad + \frac{\tau_h}{\lambda_h}\sum_{i=1}^{R_h} i\left(\rho_0(N+i)\pi(\langle 0\rangle) + \rho(N+i)\sum_{j=1}^{R_o}\pi(\langle j\rangle)\right)\\
&= \rho_0(N+R_h)\pi(\langle 0\rangle) + \rho(N+R_h)(1-\pi(\langle 0\rangle))\\
&\quad + \frac{\tau_h}{\lambda_h}\sum_{i=1}^{R_h} i(\rho_0(N+i)\pi(\langle 0\rangle) + \rho(N+i)(1-\pi(\langle 0\rangle))).
\end{aligned}
\tag{3.18}
$$

Considering Eqs. (3.15)–(3.18), we use Eq. (3.7) to calculate the approximate values of average waiting times in the buffer of unlike calls.

Let us now consider some special cases of the model under study, which are often met in the analysis of real networks. Note that to simplify the presentation the previous notation is used in partial models for their stationary distribution of state probabilities of split models and the merged model.

Model with Patient New Calls This model implies that o-calls do not leave the queue without service; in other words, $\tau_o = 0$. Then the stationary distributions inside the split models can also be calculated using Eqs. (3.10) and (3.11). However, state probabilities of the merged model are evaluated as follows in this case:

$$\pi(\langle i\rangle) = \frac{\lambda_o^*}{\mu^*}\sigma_o^{i-1}\pi(\langle 0\rangle), \quad i = 1,\ldots,R_o,$$

where

$$\mu^* = (N-g)\mu\rho(N-g),\ \sigma_o = \lambda_o/\mu^*,\ \pi(\langle 0\rangle) = \left(1 + \frac{\lambda_o^*}{\mu^*}\frac{1-\sigma_o^{R_o}}{1-\sigma_o}\right)^{-1}.$$

We find from formula (3.17) that the loss probability for o-calls is defined as follows: $P_o \approx \pi(\langle R_o\rangle)$. The other QoS parameters are calculated from the corresponding formulas.

Model with Infinite Degradation Interval This model implies that $\tau_h = 0$. In this case the stationary distributions within the split models are calculated as follows:

$$\rho_0(i) = \begin{cases} \dfrac{v^i}{i!}\rho_0(0) & \text{if } 1 \le i \le N-g, \\ \left(\dfrac{v}{v_h}\right)^{N-g}\dfrac{v_h^i}{i!}\rho_0(0) & \text{if } N-g+1 \le i \le N, \\ \left(\dfrac{v}{v_h}\right)^{N-g}\dfrac{N^N}{N!}\left(\dfrac{v_h}{N}\right)^i\rho_0(0) & \text{if } N+1 \le i \le N+R_h; \end{cases} \tag{3.19}$$

$$\rho(j) = \begin{cases} \dfrac{v_h^j}{j!}\dfrac{(N-g)!}{v_h^{N-g}}\rho(N-g) & \text{if } N-g+1 \le j \le N, \\ \left(\dfrac{v_h}{N}\right)^{j-N} v_h^g\dfrac{(N-g)!}{N!}\rho(N-g) & \text{if } N+1 \le j \le N+R_h, \end{cases} \tag{3.20}$$

where $\rho_0(0)$ and $\rho(N-g)$ are defined from appropriate normalizing conditions.

The stationary distribution of the merged model is determined similarly to Eq. (3.13). In this case, it is necessary to take into account that stationary distributions of the split models are determined from Eqs. (3.19) and (3.20). In this model, the loss probability for h-calls can be written as

$$P_h \approx \rho_0(N+R_h)\pi(\langle 0\rangle) + \rho(N+R_h)(1-\pi(\langle 0\rangle)).$$

The other QoS parameters can be found from the corresponding formulas.

Model with Patient New Calls and Infinite Degradation Interval This model is a combination of two previous models, i.e., it is assumed that $\tau_o = \tau_h = 0$. Not repeating the procedures described above, we note only that QoS metrics can be calculated by the following formulas:

$$L_o \approx \pi(\langle 0\rangle)\frac{\lambda_o^*}{\lambda_o}\sum_{i=1}^{R_o} i\sigma_o^i;$$

$$L_h \approx (a\pi(\langle 0\rangle) + b(1-\pi(\langle 0\rangle)))\sum_{i=1}^{R_h} i\left(\frac{v_h}{N}\right)^i;$$

$$P_o \approx \pi(\langle R_o\rangle);$$

$$P_h \approx (a\pi(\langle 0\rangle) + b(1-\pi(\langle 0\rangle)))\left(\frac{v_h}{N}\right)^{R_h}.$$

In what follows, the following notation is introduced:

$$a = \frac{v^{N-g}}{N!}v_h^g\rho_0(0), \quad b = \frac{(N-g)!}{N!}v_h^g\rho(N-g).$$

3.1.3 Models with Infinite Buffers

Let us now consider a model with infinite queues. Its state space is defined as follows:

$$S = \cup_{i=0}^{\infty} S_i, \tag{3.21}$$

where $S_0 = \{\boldsymbol{k} : k_1 = 0, 1, \ldots; k_2 = 0\}$, $S_i = \{\boldsymbol{k} : k_1 = N-g, N-g+1, \ldots; k_2 = i\}$, $i \geq 1$.

For simplicity, consider the model with patient new calls and infinite degradation interval for handover calls. The elements of the generating matrix of the corresponding 2-D MC can be determined similarly to Eq. (3.2). The average number of o-calls (L_o) and h-calls L_h in the queue is determined similarly to Eqs. (3.3) and (3.4), respectively, where the upper limits of summation are assumed to be infinite.

To find the stationary distribution of the model, we can use the method of 2-D generating functions. However, it involves huge computational and methodological difficulties, and on the other hand, it is nonconstructive. To overcome these difficulties, we propose to use the above approximate method of calculating the stationary state probabilities of 2-D MC. Not repeating the procedures described above, we note only that the state space splitting scheme similar to Eq. (3.1) is also used here. In this case, the stationary distribution of state probabilities of the split model with the state space S_o is defined as follows:

$$\rho_0(i)=\begin{cases}\frac{v^i}{i!}\rho_0(0) & \text{if } 1\le i\le N-g,\\ \left(\frac{v}{v_h}\right)^{N-g}\frac{v_h^i}{i!}\rho_0(0) & \text{if } N-g+1\le i\le N,\\ \frac{v^{N-g}}{N!}v_h^g\left(\frac{v_h}{N}\right)^{i-N}\rho_0(0) & \text{if } i\ge N+1,\end{cases} \tag{3.22}$$

where $\rho_0(0)$ is defined from appropriate normalizing condition, i.e.,

$$\rho_0(0)=\left(\sum_{i=0}^{N-g}\frac{v^i}{i!}+\left(\frac{v}{v_h}\right)^{N-g}\sum_{i=N-g+1}^{N}\frac{v_h^i}{i!}+\frac{v^{N-g}}{N!}v_h^{g+1}\frac{1}{N-v_h}\right)^{-1}.$$

From Eq. (3.22) we find the first ergodicity condition for the model under study: $v_h<N$.

The split models with the state space S_i are identical for all $i\ge 1$, where the birth rate is constant and equal to λ_h and the death rate in the state j is equal to $f(j)\mu$, where $j\ge N-g$. Hence, certain algebraic transformations yield that the stationary distribution of state probabilities for the split models with the state space S_i, $i\ge 1$, denoted by $\rho_i(j)$, $j\ge N-g$, can be calculated as follows:

$$\rho(j)=\begin{cases}\frac{v_h^j}{j!}\frac{(N-g)!}{v_h^{N-g}}\rho(N-g) & \text{if } N-g+1\le j\le N,\\ \left(\frac{v_h}{N}\right)^{j-N}v_h^g\frac{(N-g)!}{N!}\rho(N-g) & \text{if } j\ge N+1,\end{cases} \tag{3.23}$$

where

$$\rho(N-g)=\left(1+v_h^g(N-g)!\left(\sum_{i=N-g+1}^{N}\frac{v_h^{i-N}}{i!}+\frac{1}{N!}\frac{v_h}{N-v_h}\right)\right)^{-1}.$$

Taking into account Eqs. (3.2), (3.22), and (3.23), we find that the rates of transitions between states of the infinite-dimensional merged model are determined from the following relations:

$$q(\langle i\rangle,\langle j\rangle)=\begin{cases}\lambda_o^* & \text{if } i=0,\ j=i+1,\\ \lambda_o & \text{if } i>0,\ j=i+1,\\ \mu^* & \text{if } j=i-1,\\ 0 & \text{in other cases}.\end{cases} \tag{3.24}$$

Thus, the stationary distribution of states of the merged model is determined as the stationary distribution of states of the 1-D BDP with the rates specified by Eq. (3.25), i.e.,

$$\pi(\langle i \rangle) = \frac{\lambda_o^*}{\mu^*} \sigma_o^{i-1} \pi(\langle 0 \rangle), \, i \geq 1, \tag{3.25}$$

where

$$\pi(\langle 0 \rangle) = \left(1 + \frac{\lambda_o^*}{\mu^*} \frac{1}{1 - \sigma_o}\right)^{-1}.$$

In deriving formulas (3.25), the easy-to-verify second condition for ergodicity can be found:

$$v_o < (N - g)\rho(N - g). \tag{3.26}$$

Note 3.1 Condition (3.26) has a simple probabilistic interpretation. Since o-calls from the queue are served only if the number of channels occupied is equal to $N - g$, their total service rate is equal to $\mu(N - g)\rho(N - g)$, where $\rho(N - g)$ determines the probability of the fact that the number of channels occupied is equal to $N - g$ if there is a queue of o-calls. For the stationary mode to exist, it is required that the intensity of the incoming traffic of o-calls (λ_o) is less than the total rate of their service, and hence condition (3.26) can be found.

If both ergodicity conditions are satisfied, we find that the average number of o-calls in the queue is determined as follows:

$$L_o \approx \sum_{i=1}^{\infty} i \sum_{j=N-g}^{\infty} \rho(j)\pi(\langle i \rangle) = \sum_{i=1}^{\infty} i\pi(\langle i \rangle) = \frac{1}{(1 - \sigma_o)^2} \frac{\lambda_o^*}{\mu^*} \pi(\langle 0 \rangle). \tag{3.27}$$

The average number of h-calls in the queue is determined as follows:

$$\begin{aligned} L_h &\approx \left(\sum_{i=1}^{\infty} i\rho_0(N + i) \right) \pi(\langle 0 \rangle) + \sum_{i=1}^{\infty} i \sum_{j=1}^{\infty} \rho(N + i)\pi(\langle j \rangle) \\ &= \pi(\langle 0 \rangle) \sum_{i=1}^{\infty} i\rho_0(N + i) + (1 - \pi(\langle 0 \rangle)) \sum_{i=1}^{\infty} i\rho(N + i) \\ &= \frac{v_h^*}{(1 - v_h^*)^2} (a\pi(\langle 0 \rangle) + b(1 - \pi(\langle 0 \rangle))), \end{aligned} \tag{3.28}$$

where $v_h^* = v_h/N$.

3.1.4 Numerical Results

The developed above approximate formulas for evaluating unknown QoS metrics of models of mono-service networks allow analyzing them easily for arbitrary volumes of buffer stores for unlike calls.

First, let us consider the results of numerical experiments for the general model with finite buffers, i.e., for the model where $\tau_o \neq 0$ and $\tau_h \neq 0$. We will analyze the behavior of QoS metrics with respect to variations in the parameter g for fixed values of other parameters of the model. The corresponding curves for a hypothetical model are shown in Figs. 3.2, 3.3, and 3.4, where the initial data were taken as follows: $N=10$; $\lambda_o=0.2$; $\lambda_h=2.6$; $\mu=5$; $\tau_o=0.1$ and $\tau_h=0.2$. As is seen from

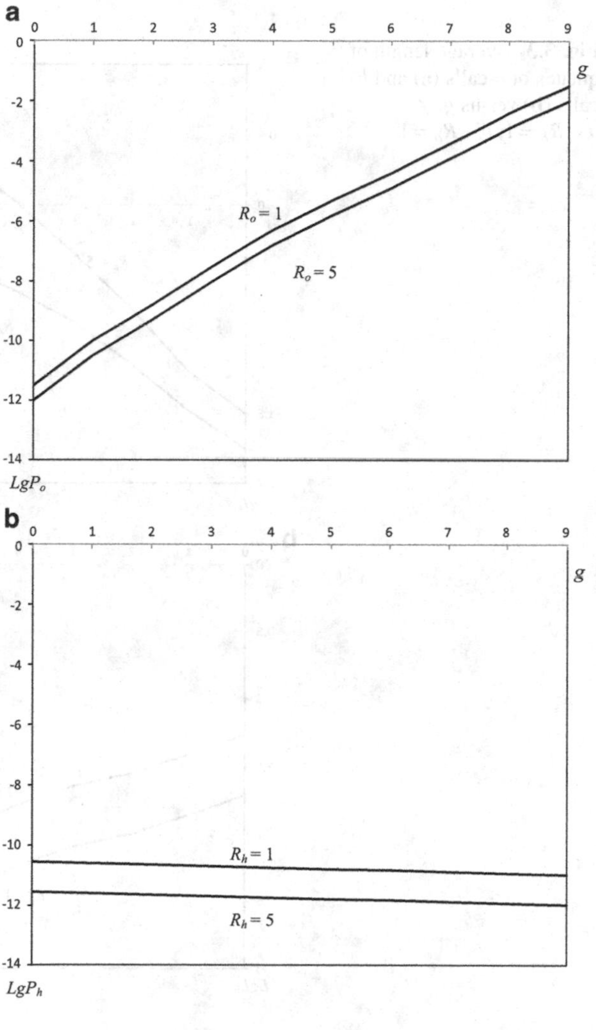

Fig. 3.2 Blocking probabilities of o-calls (**a**) and h-calls (**b**) versus g, $a-R_h=1$; $b-R_o=5$

these curves, an increase in the parameter g increases the loss probability of o-calls (see Fig. 3.2a) and decreases the loss probability of h-calls (see Fig. 3.2b). This can be explained by the fact that as the number of guard channels increases, the chances of o-calls to access the channels decrease, and vice versa, the chances of h-calls to access the channels increase. As one would expect, an increase in the volume of the buffer for calls of each type (for a fixed buffer volume for calls of other type) reduces their loss probability (see Fig. 3.2). Note also that growth of the rate of any flow increases the loss probability.

The average queue length of o-calls grows with increase in the number of guard channels (see Fig. 3.3a), and the corresponding parameter for h-calls decreases with respect to the parameter specified (see Fig. 3.3b). At the same time, both functions are increasing with respect to the size of the corresponding buffer. Average waiting times for unlike calls in the queue have a similar form (see Fig. 3.4).

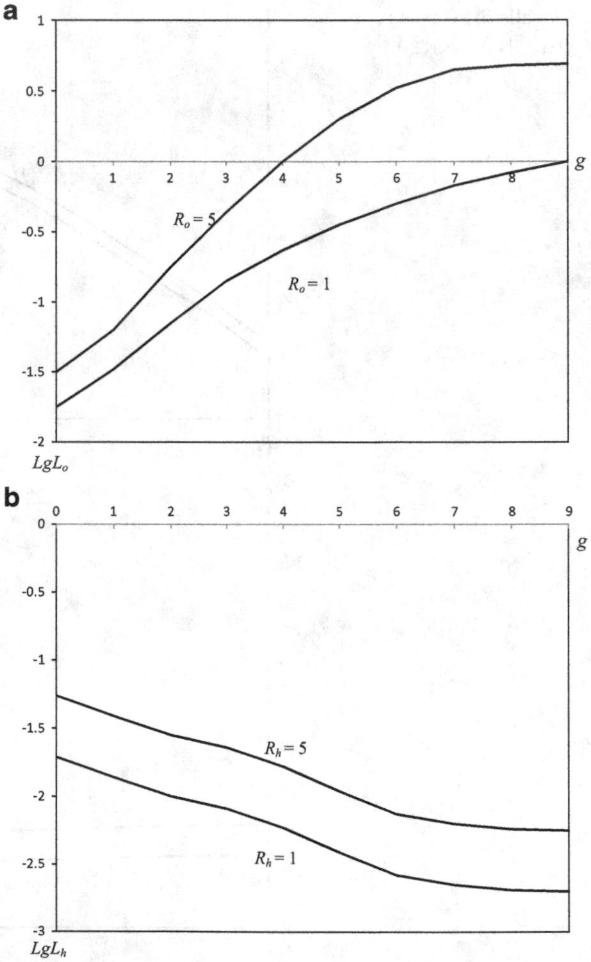

Fig. 3.3 Average length of queues of o-calls (**a**) and h-calls (**b**) versus g, $a - R_h = 1$; $b - R_o = 1$

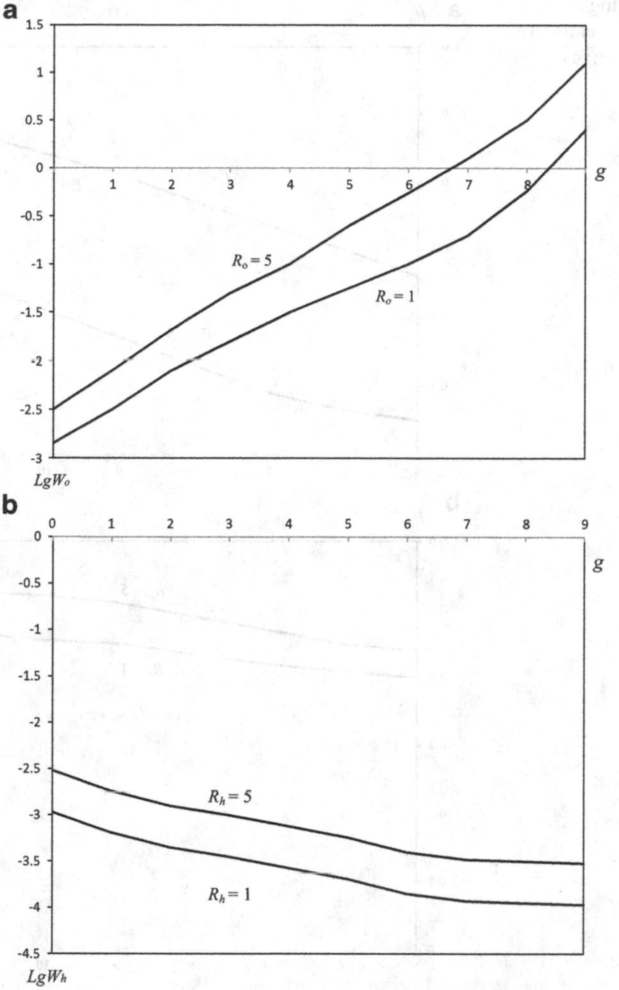

Fig. 3.4 Average waiting times of *o*-calls (**a**) and *h*-calls (**b**) versus g, $a - R_h = 1$; $b - R_o = 1$

The behavior of QoS metrics with respect to variations in the intensities of incoming traffic for fixed values of other parameters was also analyzed. Note that these studies are important since in practice the rates of incoming traffic are determined with some errors, and their values vary with time. Therefore, of great interest is the analysis of invariance (or weak variability) intervals of QoS metrics of the model with respect to variations in the intensity of incoming traffic for fixed values of other parameters. The corresponding results are shown in Figs. 3.5, 3.6, and 3.7.

Another field of studies was the accuracy of the formulas developed to evaluate the approximate values (AVs) of QoS parameters of the model under study.

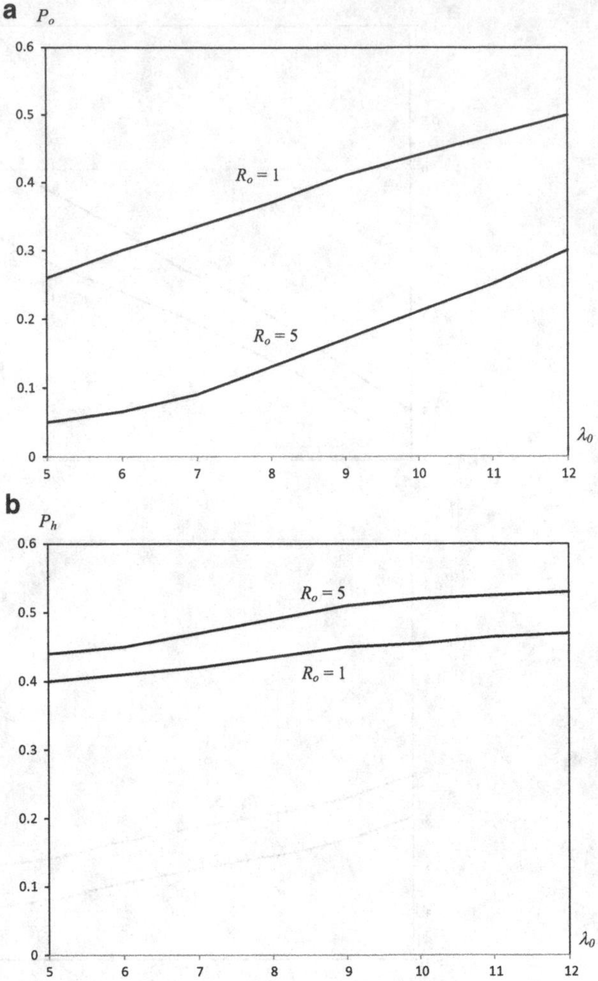

Fig. 3.5 Blocking probabilities of o-calls (**a**) and h-calls (**b**) versus λ_o

As exact values (EVs) of these quantities, their values calculated using the approach stated in [13] were used. As indicated above, such an approach allows studying QoS parameters of the model only for small buffer stores. For the above-indicated initial data and for $R_o = R_h = 1$, the corresponding AVs and EVs are compared in

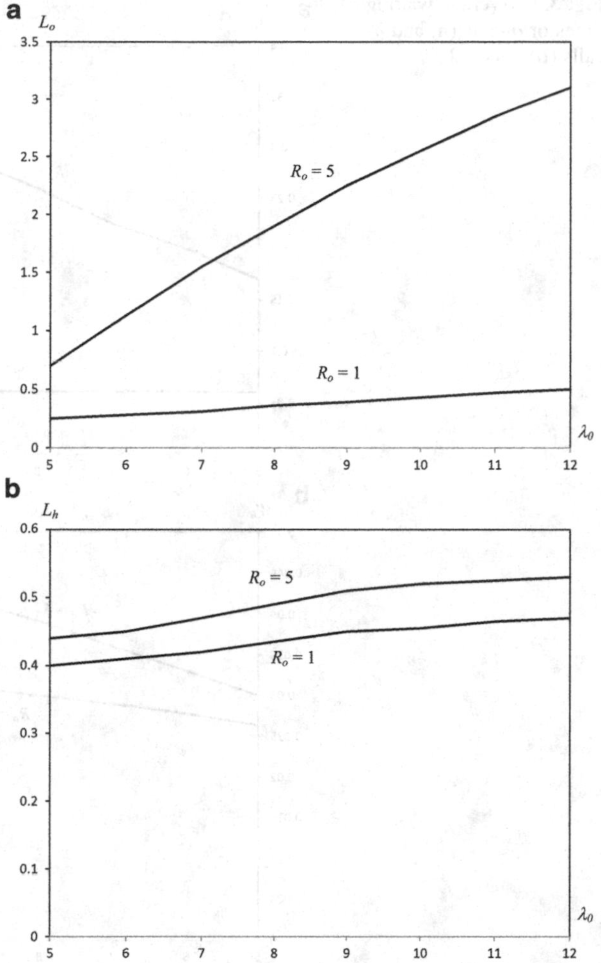

Fig. 3.6 Average length of queues of o-calls (**a**) and h-calls (**b**) versus λ_o

Tables 3.1 and 3.2. As follows from the tables, the approximate formulas derived are sufficiently accurate.

Let us now consider some results of numerical experiments for a model with infinite buffer stores. The initial data for a hypothetical model were taken as follows: $N = 30$, $\lambda_h = 12$ and $\mu = 1$. For $\lambda_o = 2$ and $\lambda_o = 4$, the ergodicity condition

Fig. 3.7 Average waiting times of *o*-calls (**a**) and *h*-calls (**b**) versus λ_o

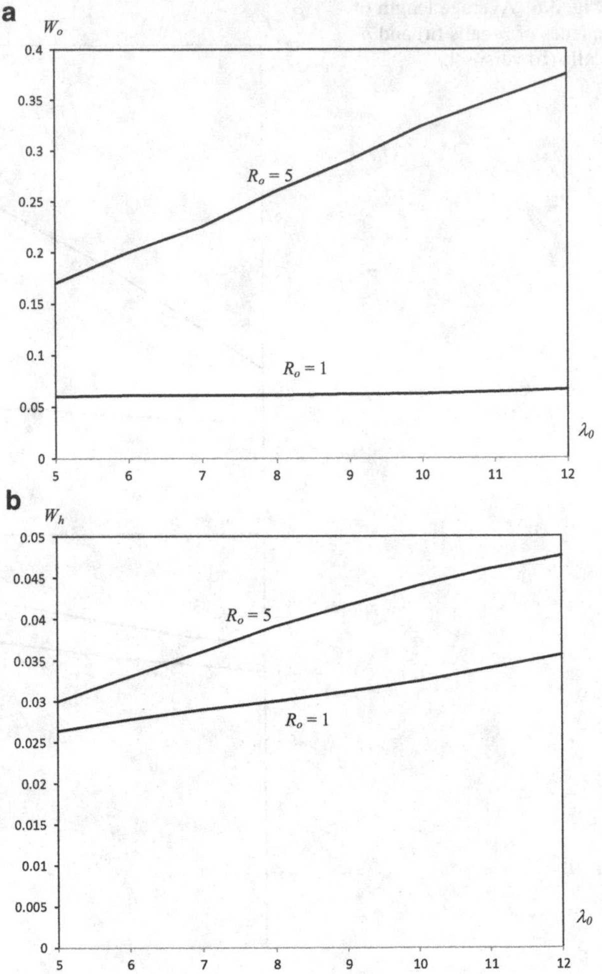

(3.27) is satisfied for $0 \leq g \leq 15$ and for $0 \leq g \leq 18$, respectively; therefore, these ranges of the parameter g are specified for the corresponding curves (Fig. 3.8). As is seen from Fig. 3.8, an increase in the parameter g leads to an increase in the function L_o but reduces the function L_h.

These results have quite logical explanation since as the number of guard channels increases, the chances of *h*-calls for direct access to channels grow;

Table 3.1 Comparison for o-calls in mono-service model with finite buffers

	P_o		L_o		W_o	
g	EV	AV	EV	AV	EV	AV
0	3.16236E-12	2.43785E-12	1.62726E-02	1.62730E-02	1.65440E-03	1.65453E-03
1	6.30767E-11	7.86103E-11	3.73811E-02	3.73832E-02	3.88478E-03	3.88485E-03
2	1.15364E-09	2.44697E-09	7.42411E-02	7.42422E-02	8.02593E-03	8.02599E-03
3	1.91132E-08	5.78531E-08	1.34232E-01	1.34237E-01	1.55284E-02	1.55296E-02
4	2.84055E-07	3.14851E-07	2.24514E-01	2.24518E-01	2.90355E-02	2.90361E-02
5	3.74555E-06	3.77654E-06	3.49392E-01	3.49395E-01	5.39924E-02	5.39931E-02
6	4.33084E-05	4.35672E-05	5.06623E-01	5.06628E-01	1.03750E-01	1.03757E-01
7	4.34859E-04	4.35532E-04	6.82404E-01	6.82407E-01	2.19583E-01	2.19589E-01
8	3.80973E-03	3.80652E-03	8.45889E-01	8.45890E-01	5.80761E-01	5.80763E-01
9	3.17847E-02	3.17850E-02	9.54261E-01	9.54271E-01	2.63631E+00	2.63636E+00

Table 3.2 Comparison for h-calls in mono-service model with finite buffers

	P_h		L_h		W_h	
g	EV	AV	EV	AV	EV	AV
0	2.67415E-11	4.04561E-11	2.12586E-02	2.12673E-02	1.08633E-03	1.08641E-03
1	2.48432E-11	3.77538E-11	1.48829E-02	1.48832E-02	7.55501E-04	7.55512E-04
2	2.30828E-11	3.72416E-11	1.04747E-02	1.04768E-02	5.29343E-04	5.29367E-04
3	2.14511E-11	3.68643E-11	7.47977E-03	7.47985E-03	3.76848E-04	3.76853E-04
4	1.99403E-11	2.99853E-11	5.45992E-03	5.45979E-03	2.74515E-04	2.74528E-04
5	1.85443E-11	2.92375E-11	4.10648E-03	4.10661E-03	2.06187E-04	2.0618E-04
6	1.72599E-11	2.83243E-11	3.21075E-03	3.21112E-03	1.61061E-04	1.61059E-04
7	1.60904E-11	2.73051E-11	2.63707E-03	2.63853E-03	1.32213E-04	1.32235E-04
8	1.50534E-11	2.61631E-11	2.30184E-03	2.30195E-03	1.15367E-04	1.15372E-04
9	1.41935E-11	2.58932E-11	2.14829E-03	2.14852E-03	1.07654E-04	1.07663E-04

however, at the same time, the chances of o-calls for direct access to channels decrease and they are forcedly queued. Note that the average number of o-calls in the queue grows with a greater velocity than the average number of h-calls in the queue (see Fig. 3.8b). As is seen from these curves, there is no need for the system to organize an infinite buffer to wait for unlike calls since average lengths of queues of unlike calls are rather short. For example, if buffers of sizes 5 and 3 are organized in a hypothetical network under study to wait for o- and h-calls in queue, respectively, they will appear to be sufficient to handling the unlike calls, their loss probability being within comprehensible limits. In other words, in each specific case, there are possibilities to study the choice of necessary volumes of buffer stores to satisfy specified constraints for QoS metrics of the network.

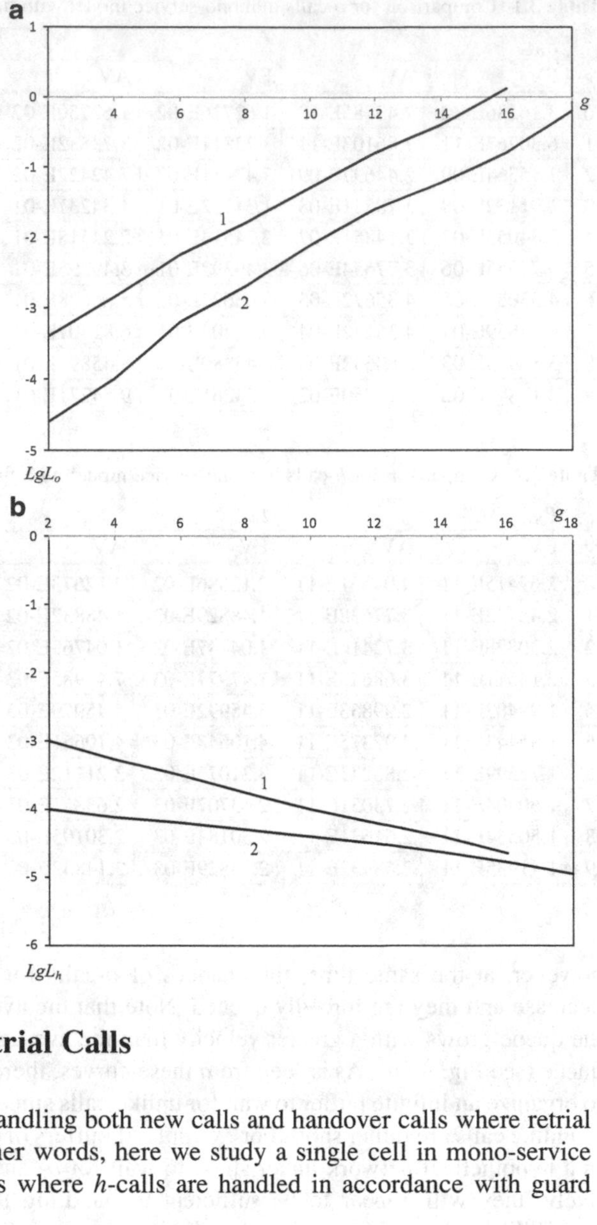

Fig. 3.8 Average length of queues of o-calls (**a**) and h-calls (**b**) versus g in the model with infinite queues, $1 - \lambda_o = 4$; $2 - \lambda_o = 2$

3.2 Models with Retrial Calls

Now consider a cell with handling both new calls and handover calls where retrial phenomenon occurs. In other words, here we study a single cell in mono-service network with retrial o-calls where h-calls are handled in accordance with guard channel scheme [26].

Note that if upon arrival of h-call there is at least one free channel, it is admitted; otherwise it is forcedly terminated. When o-call's arrival finds g or more channels being busy, this o-call is blocked and enters the retrial group with probability H_o or leaves the system forever with probability $1 - H_o$. The blocked calls in the retrial group try to redial after retrial time whose distribution is exponential with rate η. The blocked o-calls in the retrial group after an unsuccessful retrial return to retrial group with probability H_1 or leave the system with probability $1 - H_1$. Channel holding time of both types of calls is exponential with the same parameter μ. The capacity of the retrial group (R) might be either finite or infinite.

3.2.1 Exact Method

First we assume that the capacity of the retrial group is infinite, i.e., $R=\infty$. The state of a cell at an arbitrary time can be described by a two-dimensional vector $\boldsymbol{k}=(k_1, k_2)$, where k_1 indicates the total number of blocked o-calls in the retrial group and k_2 is the number of calls in service. The state space of the system is defined as follows:

$$S=\{\boldsymbol{k}:\ k_1=0,1,\ldots;\ k_2=0,1,\ldots,N\}. \tag{3.29}$$

On the basis of the adopted access scheme, we can conclude that the nonnegative elements of Q-matrix of the given 2-D MC are given by (see Fig. 3.9)

$$q(\boldsymbol{k},\boldsymbol{k}')=\begin{cases}\lambda & \text{if } k_2<g,\ \boldsymbol{k}'=\boldsymbol{k}+\boldsymbol{e}_2,\\ \lambda_h & \text{if } k_2\geq g,\ \boldsymbol{k}'=\boldsymbol{k}+\boldsymbol{e}_2,\\ k_1\eta & \text{if } k_2<g,\ \boldsymbol{k}'=\boldsymbol{k}-\boldsymbol{e}_1+\boldsymbol{e}_2,\\ k_1\eta(1-H_1) & \text{if } k_2\geq g,\ \boldsymbol{k}'=\boldsymbol{k}-\boldsymbol{e}_1,\\ \lambda_o H_o & \text{if } k_2\geq g,\ \boldsymbol{k}'=\boldsymbol{k}+\boldsymbol{e}_1,\\ k_2\mu & \text{if } \boldsymbol{k}'=\boldsymbol{k}-\boldsymbol{e}_2,\\ 0 & \text{in other cases.}\end{cases} \tag{3.30}$$

State probabilities satisfy the following SGBE which is constructed on the basis of relations (3.30):

For case $k_2<g$,

$$\begin{aligned}(\lambda+k_1\eta+k_2\mu)p(\boldsymbol{k}) &= \lambda p(\boldsymbol{k}-\boldsymbol{e}_2)+(k_2+1)\mu p(\boldsymbol{k}+\boldsymbol{e}_2)\\ &\quad+(k_1+1)\eta p(\boldsymbol{k}+\boldsymbol{e}_1-\boldsymbol{e}_2).\end{aligned} \tag{3.31}$$

For case $g\leq k_2\leq N$,

$$\begin{aligned}&(\lambda_h+\lambda_o H_o+k_1\eta(1-H_1)+k_2\mu)p(\boldsymbol{k})\\ &=\lambda_o H_o p(\boldsymbol{k}-\boldsymbol{e}_1)+(k_2+1)\mu p(\boldsymbol{k}+\boldsymbol{e}_2)\\ &\quad+(k_1+1)\eta(1-H_1)p(\boldsymbol{k}+\boldsymbol{e}_1).\end{aligned} \tag{3.32}$$

We add normalizing condition to SGBE (3.31), (3.32):

$$\sum_{\boldsymbol{k}\in S} p(\boldsymbol{k})=1. \tag{3.33}$$

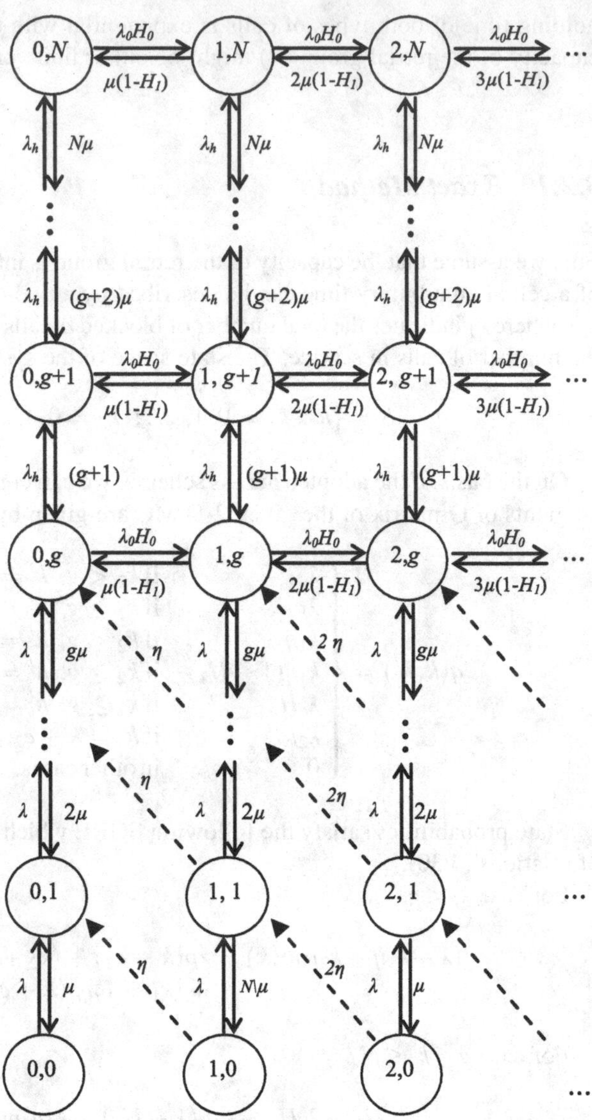

Fig. 3.9 State diagram of the model with retrial calls and infinite capacity of retrial group

The main QoS metrics of special interest are:

(i) The probability that an o-call does not receive service on its first attempt is

$$P_o = \sum_{k_1=0}^{\infty} \sum_{k_2=g}^{N} p(k_1, k_2). \tag{3.34}$$

(ii) Loss probability of h-calls is

$$P_h = \sum_{k_1=0}^{\infty} p(k_1, N). \tag{3.35}$$

(iii) The probability of a random retrial o-call being blocked is

$$P_r = \sum_{k_1=1}^{\infty} \sum_{k_2=g}^{N} p(k_1, k_2). \tag{3.36}$$

(iv) The average number of an o-call in retrial group is

$$L_r = \sum_{k_1=1}^{\infty} \sum_{k_2=0}^{N} k_1 p(k_1, k_2). \tag{3.37}$$

(v) The average rate of the retrial o-call blocking is

$$R_r = \sum_{k_1=1}^{\infty} \sum_{k_2=g}^{N} k_1 p(k_1, k_2). \tag{3.38}$$

An analytical solution of stationary distribution of multi-server retrial queue with H_0, $H_1 > 0$ is not known yet. So we need an approximation method.

3.2.2 Approximate Method

For accurate application of phase merging algorithms, it is assumed that transition intensities between states within one column are considerably greater than those of different columns (see Fig. 3.9). Note that this assumption is realistic particularly in the cellular networks where incoming or servicing intensities of one type of calls exceed greatly those of different type. In other words, we need to assume that $\lambda_h \gg \lambda_o H_o$, $\mu \gg \eta$ and $\lambda_h \gg \eta(1 - H_1)$. As it was mentioned above (see Sect. 3.1.1), this assumption is not extraordinary for wireless networks, since this is a regime that commonly occurs in microcells, in which mobile users have high mobility and short duration calls [6, 10].

Consider the following splitting of state space (3.29):

$$S = \cup_{i=0}^{\infty} S_i, S_i \cap S_j = \varnothing, \text{if } i \neq j, \tag{3.39}$$

where $S_i = \{\mathbf{k} \in S : k_1 = i, 0 \le k_2 \le N\}$.

Based on splitting (3.39) the sets S_i are then united into individual merged states $\langle i \rangle$, and the following merged function with the domain (3.29) is introduced (see Fig. 3.9):

$$U(\boldsymbol{k}) = \langle i \rangle \text{ if } \boldsymbol{k} \in S_i, \ i = 0, 1, \ldots .$$

State probabilities within classes S_i are denoted by $\rho_i(j)$, $i = 0, 1, \ldots, j = 0, 1, \ldots, N$. These probabilities are independent on index i, and they coincide with state probabilities of the model $M/M/N/N$ with constant service rate μ and state-dependent arrival rates which are determined as

$$\lambda_j = \begin{cases} \lambda_o + \lambda_h & \text{if } 0 \le j \le g-1, \\ \lambda_h & \text{if } g \le j \le N-1. \end{cases}$$

Hence the indicated state probabilities are

$$\rho_i(j) = \begin{cases} \dfrac{v^j}{j!}\rho_i(0) & \text{if } 1 \le j \le g, \\ \left(\dfrac{v}{v_h}\right)^g \dfrac{v_h^j}{j!}\rho_i(0) & \text{if } g < j \le N, \end{cases} \tag{3.40}$$

where $v = (\lambda_o + \lambda_h)/\mu$, $v_h = \lambda_h/\mu$,
$\rho_i(0) = \left(\sum_{j=0}^{g} (v^j/j!) + (v/v_h)^g \sum_{j=g+1}^{N} \left(v_h^j/j!\right)\right)^{-1}$.

Since quantities $\rho_i(j)$, $i = 0, 1, \ldots, j = 0, 1, \ldots, N$ do not depend on i below in the notation $\rho_i(j)$, this index is omitted. Then from Eqs. (3.30) and (3.40), we conclude that nonnegative elements of Q-matrix are

$$q(\langle i \rangle, \langle j \rangle) = \begin{cases} \lambda_o H_0 \sum_{n=g}^{N} \rho(n) & \text{if } i \ge 0, \ j = i+1, \\ i\eta \left(\sum_{n=0}^{g-1} \rho(n) + (1 - H_1) \sum_{n=g}^{N} \rho(n) \right) & \text{if } i \ge 1, \ j = i-1, \\ 0 & \text{in other cases.} \end{cases} \tag{3.41}$$

Thus, the stationary distribution of a merged model $\pi(\langle i \rangle)$, $i = 0, 1, \ldots$ is defined as the stationary distribution of classical model $M/M/\infty$ with load $v_\infty = \lambda_\infty/\mu_\infty$, where

$$\lambda_{\infty} = \lambda_o H_0 \sum_{n=g}^{N} \rho(j), \quad \mu_{\infty} = \eta \left(\sum_{n=0}^{g-1} \rho(n) + (1 - H_1) \sum_{n=g}^{N} \rho(n) \right).$$

In other words, the stationary distribution of a merged model is determined as

$$\pi(\langle i \rangle) = \frac{v_{\infty}^i}{i!} e^{-v_{\infty}}, \; i = 0, 1, 2, \ldots. \tag{3.42}$$

By using Eqs. (3.40) and (3.42) state probabilities of the initial 2-D MC is found as follows:

$$p(k_1, k_2) \approx \rho(k_2) \pi(\langle k_1 \rangle).$$

After some mathematical transformation the following approximate formulas to calculate the QoS metrics of the model with infinite capacity of retrial group are found:

$$P_o \approx 1 - \rho(0) \sum_{n=0}^{g-1} \frac{v^n}{n!} \tag{3.43}$$

$$P_{\mathrm{r}} \approx (1 - e^{-v_{\infty}}) P_o \tag{3.44}$$

$$P_h \approx \rho(0) \frac{v^g v_h^{N-g}}{N!} \tag{3.45}$$

$$L_r \approx v_{\infty} \tag{3.46}$$

$$R_{\mathrm{r}} \approx v_{\infty} P_o \tag{3.47}$$

The developed approximate approach might be used for the model with finite capacity of retrial group, i.e., for case $R < \infty$. Note that in this case the exact approach can be successfully used for the model with moderate size of retrial group.

Dropping well-known steps in solution of this problem, below are given the final formulas for approximate calculation of QoS metrics of the model with finite capacity of retrial group. Note that in this case the stationary distribution of a merged model is the same as the classical Erlang's model *M*/*M*/*R*/*R* with load v_{∞} Erl. The QoS metrics P_o and P_h are determined by the formulas (3.43) and (3.44), respectively. In other words, these QoS metrics are independent of the capacity of retrial group. This fact can be easily explained if one takes into account the accepted above assumptions concerning the ratio of load characteristics of the model. However to calculate the remains of QoS metrics, the following approximate formulas are obtained:

$$P_{\mathrm{r}} \approx (1 - \pi(\langle 0 \rangle))P_o \tag{3.48}$$

$$L_{\mathrm{r}} \approx \pi(\langle 0 \rangle)\sum_{i=1}^{R} \frac{v_{\infty}^{i}}{(i-1)!} \tag{3.49}$$

$$R_{\mathrm{r}} \approx L_{\mathrm{r}}P_o \tag{3.50}$$

3.2.3 Numerical Results

Consider some numerical results for model with retrial calls. In both cases $R=5$ and $R=\infty$, the initial data is chosen as follows: $N=20$, $\lambda_h=20, \lambda_o=2, \mu=5, \eta=0.5$, $H_0=0.9, H_1=0.6$.

In Fig. 3.10 the dependency of both functions P_o and P_h on the parameter g is shown. Note that for given input data values of indicated functions in both cases $R=5$ and $R=\infty$ are almost same. As it was expected, function P_o is decreasing function versus g, since increase of the value of given parameter leads to increase of the access chances of o-calls to channels; however, at the same time increase of the value of parameter g leads to decrease of the access chances of h-calls to channels, i.e., function P_h is increasing function versus g.

Figure 3.11 demonstrates the dependency of function P_{r} on the parameter g in both cases $R=5$ and $R=\infty$. It is seen from this figure that the blocking probability of a random retrial o-call in both cases is decreasing function. This fact is also excepted one, since, as it is mentioned above, increase of the value of parameter g leads to increase of the access chances of o-calls to channels.

Dependency of function L_{r} on the parameter g is shown in Fig. 3.12. The average number of an o-call in retrial group in the case $R=5$ is almost constant (i.e., does not depend on the parameter g), while in the case $R=\infty$ it has high decreasing rate. The curve for an average rate of blocking of the retrial o-calls (Fig. 3.13) has precisely the same form as a curve for blocking probability of retrial o-calls (see Fig. 3.11).

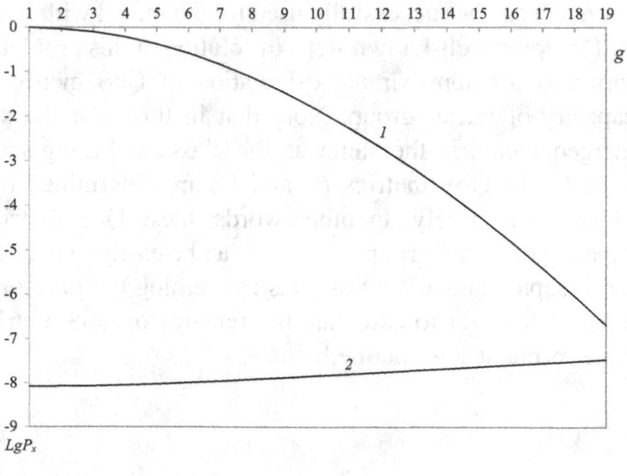

Fig. 3.10 Blocking probability of new and handover calls versus g; $1 - P_o$; $2 - P_h$

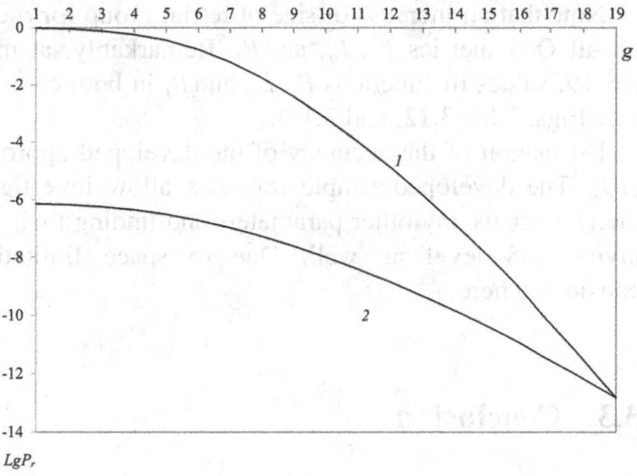

Fig. 3.11 Blocking probability of retrial o-calls versus g; $1-R=\infty$; $2-R=5$

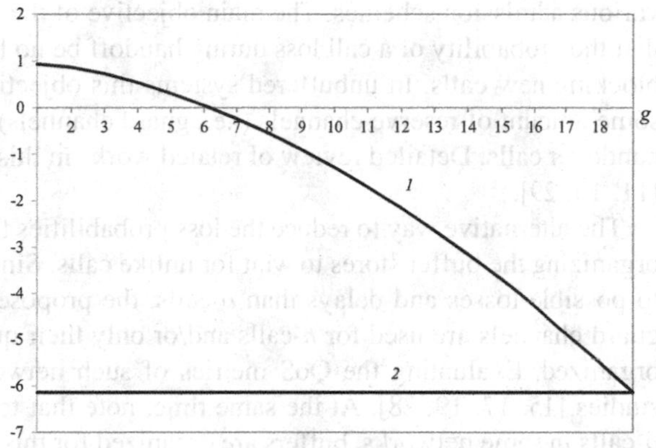

Fig. 3.12 Average number of an o-calls in retrial group; $1-R=\infty$; $2-R=5$

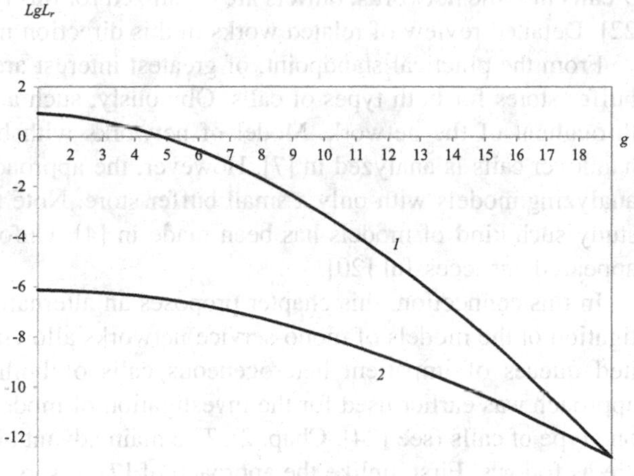

Fig. 3.13 Average rate of blocking of the retrial o-calls; $1-R=\infty$; $2-R=5$

Note that an increase of size of retrial group for o-calls has no beneficial effect on all QoS metrics P_r, L_r, and R_r. Remarkably, at maximal value of parameter $g = 19$, values of functions P_r, L_r, and R_r in both cases $R = 5$ and $R = \infty$ are same (see Figs. 3.11, 3.12, and 3.13).

Estimation of the accuracy of the developed approximate formulas is given in [10]. The developed simple formulas allow investigating the behavior of QoS metrics versus any other parameters and finding their desired values to satisfy the given QoS level as well. Due to space limitation these results are not considered here.

3.3 Conclusion

There is a very large literature published in mono-service cellular networks with various admission schemes. The main objective of the early investigations requires that the probability of a call loss during handoff be no larger than the probability of blocking new calls. In unbuffered systems this objective is achieved by reserving some amount of reserve channels (i.e., guard channels) which are available only to handover calls. Detailed review of related works in this direction might be found in [11, 13, 29].

The alternative way to reduce the loss probabilities for new and handover calls is organizing the buffer stores to wait for unlike calls. Since h-calls are more sensitive to possible losses and delays than o-calls, the proposed schemes often imply that guard channels are used for h-calls and/or only their queues in the base station are organized. Evaluating the QoS metrics of such networks is the subject of many studies [15, 17, 19, 28]. At the same time, note that to compensate for chances of o-calls in some networks, buffers are organized for this type of calls as well [14, 18, 22]. Detailed review of related works in this direction might be found in [1, 4, 5].

From the practical standpoint, of greatest interest are models of networks with buffer stores for both types of calls. Obviously, such a scheme increases the total throughput of the network. Model of networks with buffering of both new and handover calls is analyzed in [7]. However, the approach proposed therein allows analyzing models with only a small buffer store. Note that another attempt to the study such kind of models has been made in [4]. Unfortunately, this attempt has appeared unsuccessful [20].

In this connection, this chapter proposes an alternative approach for the investigation of the models of mono-service networks allowing both limited and unlimited queues of impatient heterogeneous calls of both types [21, 25]. Such an approach was earlier used for the investigation of models allowing queues of only one type of calls (see [24], Chap. 2). The main advantages of the proposed method are as follows. First, unlike the approach of [7], it serves to study not only models with small dimensions of their buffer storage but also with large dimensions of their buffer storage; second, at the use of this approach, simple analytical formulas for the calculation of the sought QoS metrics of the studied networks have been successfully developed; third, the method developed makes it possible to solve the important problems of determining rational sizes of buffer stores and/or the number of guard channels to optimize the desired QoS metrics of a network.

The problem of development of effective numerical algorithms for calculation of retrial queueing models of wireless networks is a subject for many researches. Detailed list of works in this direction might be found in [2]. Note that the main mathematical tool to investigate such kind of models is matrix analytical method. In order to use this method, the maximum number of blocked call who can retry to access the servers is restricted by some threshold parameter. So, by restricting finite capacity of retrial group (see [8, 9]) or by allowing a fixed number of blocked calls in the infinite capacity of retrial group to be able to retry (see [3, 16, 23]), the matrix analytical method can be applied to the multidimensional Markov chain. In [27], the condition of the existence of stationary mode in model of retrial queue with variable arrival rate is found, and the approximate approach to calculate its state probabilities is proposed. Recently model of retrial queue with the fractional guard channel policy is applied in mono-service cellular networks [12]. In this chapter, an algorithmic method is applied to cell's model to find approximate stationary distribution and then to obtain its performance measures. As author's knowledge is concerned, analytical solution of stationary distribution of multi-server retrial queue with positive retrial probabilities is not known yet. Developed here is the simple approximation method based on the results of the paper [10].

References

1. Abdulova V, Aybay I (2011) Predictive mobile-oriented channel reservation schemes in wireless cellular networks. Wirel Netw 17:149–166
2. Artalejo JR (2011) Accessible bibliography on retrial queues: Progress in 2000-2009. Math Comput Model 51:1071–1081
3. Artalejo JR, Orlovsky DS, Dudin AN (2005) Multi-server retrial model with variable number of active servers. Comput Ind Eng 48(2):273–288
4. Baloch RA, Awan I, Min G (2010) A mathematical model for wireless channel allocation and handoff schemes. Telecommun Syst 45:275–287
5. Bozkurt A, Akdeniz R, Ucar E (2010) Call admission control jointly with resource reservation in cellular wireless networks. EURASIP J Wirel Commun Netw 2010:Article ID 740575, 10 pages
6. Casares-Giner V (2001) Integration of dispatch and interconnect traffic in a land mobile trunking system. Waiting time distributions. Telecommun Syst 10:539–554
7. Chang CJ, Su TT, Chiang YY (1994) Analysis of a cutoff priority cellular radio system with finite queuing and reneging/dropping. IEEE/ACM Trans Netw 2(2):166–175
8. Choi BD, Chang Y (1999) Single server retrial queues with priority calls. Math Comput Model 30(3):7–32
9. Choi BD, Chang Y (1999) MAP1, MAP2/M/c retrial queue with the retrial group of finite capacity and geometric loss. Math Comput Model 3(3):99–113
10. Choi BD, Melikov AZ, Velibekov AM (2008) A simple numerical analysis of joint probabilities of calls in service and calls in the retrial group in a picocell. Appl Comput Math 7 (1):25–34

11. Das Bit S, Mitra S (2003) Challenges of computing in mobile cellular environment—a survey. Comput Commun 26:2090–2105
12. Do TV (2011) Solution for a retrial queueing problem in cellular networks with the fractional guard channel policy. Math Comput Model 53:2059–2066
13. Du W, Lin L, Jia W, Wang G (2005) Modeling and performance evaluation of handover service in wireless networks. In: Lu X, Zhao W (eds) Lectures Notes in Computer Science, vol 3619. Springer, Heidelberg, pp 229–238
14. Guerin R (1988) Queuing-blocking system with two arrival streams and guard channels. IEEE Trans Commun 36(2):153–163
15. Hong D, Rapoport SS (1986) Traffic model and performance analysis of cellular mobile radio telephones systems with prioritized and non-prioritized handoff procedures. IEEE Trans Veh Technol 35(3):77–92
16. Kim CS, Klimenok VI, Dudin AN (2014) Analysis and optimization of guard channel policy in cellular mobile networks with account of retrials. Comput Oper Res 43:181–190
17. Kim CS, Melikov AZ, Ponomarenko LA (2007) Two-dimensional models of wireless cellular networks with infinite queues of handover calls. J Autom Inform Sci 39(12):25–41
18. Kim CS, Ponomarenko LA, Melikov AZ (2008) Numerical approach to analysis of channel assignment schemes in mobile cellular networks. J Korea Manag Eng Soc 13(2):29–39
19. Louvros S, Pylarinos J, Kotsopoulos S (2007) Mean waiting time analysis in finite storage queues for wireless cellular networks. Wirel Personal Commun 40:145–155
20. Melikov AZ (2014) Comment on "A mathematical model for wireless channel allocation and handoff schemes". Telecommun Syst (In press)
21. Melikov AZ, Fattakhova MI (2010) Simulation of wireless cellular networks with impatient calls. Automat Contr Comput Sci 44(1):31–41
22. Melikov AZ, Velibekov AM (2009) Computational procedures for analysis of two channels assignment schemes in wireless cellular networks. Automat Contr Comput Sci 43(2):96–103
23. Mushko VV, Klimenok VI, Ramakrishnan KO, Krishnamoorthy A, Dudin AN (2006) Multi-server queue with addressed retrials. Ann Oper Res 141(1):283–301
24. Ponomarenko L, Kim CS, Melikov A (2010) Performance analysis and optimization of multi-traffic on communication networks. Springer, Heidelberg
25. Ponomarenko LA, Melikov AZ, Fattakhova MI (2010) Approximate evaluation of models of microcellular wireless networks with queues of unlike calls. Cybern Syst Anal 46(3):460–471
26. Servi LD (2002) Algorithmic solutions to two-dimensional birth-death processes with application to capacity planning. Telecommun Syst 21(2–4):205–212
27. Usar IY, Lebedev EA (2013) Retrial queueing systems with variable arrival rate. Cybern Syst Anal 49(3):457–464
28. Yoon CH, Un CK (1993) Performance of personal portable radio telephone systems with and without guard channels. IEEE J Sel Area Commun 11(6):911–917
29. Zeng H, Chlamtac I (2003) Adaptive guard channel allocation and blocking probability estimation in PCS networks. Comput Netw 43:163–176

Chapter 4
Algorithmic Methods for Analysis of Integral Cellular Networks

In the previous chapter, the second-generation cellular networks which are primarily concerned with single service class (voice traffic) communications were investigated. However, future generation wireless networks need to incorporate multimedia services (e.g., voice, video, data, etc.). In such networks all traffics (type of services) can be broadly divided into two types of real-time (such as voice, video transmission) and non-real-time (such as data transmission) services. In this chapter, models of integral cellular networks with real-time (RT) and non-real-time (NRT) calls are investigated. For easy reference here we define the RT calls as voice calls, while NRT calls are considered as data calls. In such networks four types of calls are distinguished: handover voice calls (hv-calls), new (or originating) voice calls (ov-calls), handover calls of data (hd-calls), and new (or originating) calls of data (hd-calls). The importance of these calls decreases in the above order. Voice calls (v-calls) are handled according to the pure loss scheme (i.e., the calls not accepted at the time of arrival are lost), and data calls (d-calls) are relatively tolerant to possible delays, i.e., they can wait in a queue of finite or infinite length. The access of heterogeneous voice calls is controlled by means of two-parameter state-dependent strategies, which limits the access on new and handover voice calls. Two schemes for buffering of data calls have been considered. In the first scheme only handover data calls can form a queue of finite and infinite length, while in the second one both new and handover data calls might be waiting in queue. In the second scheme the access of new data calls to buffer is restricted by means of state-dependent reservation scheme. Both methods for exact and approximate evaluation of the QoS metrics in such networks are developed. There has been also an investigation of the problems of choosing the required values of parameters for the introduced access strategies satisfying the given quality levels of servicing the heterogeneous calls. The results of comparative analysis of QoS metrics of the model under different access schemes have been presented. Results of numerical experiments are presented.

A. Melikov, L. Ponomarenko, *Multidimensional Queueing Models in Telecommunication Networks*, DOI 10.1007/978-3-319-08669-9_4

4.1 Models with Buffering of Handover Non-Real-Time Calls

In this section we consider the models in which only buffering of hd-calls is admitted and for access control several threshold parameters are introduced. The indicated parameters restrict the access of heterogeneous calls (except for hv-calls) depending on the state of the cell. Consideration is given to an isolated cell of homogeneous multiservice wireless cellular network in which voice calls and data packages (further just data calls) are being processed. Network homogeneity means that traffics in its different cells are statistically identical and, hence, the network study on a cell level is proper. Note that in networks of microcell structure (i.e., with relatively small geometrical sizes of cells), the assumption on statistical identity of heterogeneous traffics is almost always fulfilled.

We consider two state-dependent schemes to restrict the access of heterogeneous voice calls to channels. In the first scheme the decision on access of heterogeneous v-calls is taken on the basis of a general number of such kind of calls in a cell (cutoff scheme), while in the second scheme the indicated decision is based on the total number of busy channels of the cell (guard channel scheme).

4.1.1 Various Access Schemes to Channels

The network applies a fixed scheme of channel distributions among its cells and each cell has $N > 1$ radio channels. Channels are used jointly by Poisson flows of hv-, ov-, hd-, and od-calls. The intensity of x-calls is denoted by $\lambda_x, x \in \{\text{hv}, \text{ov}, \text{hd}, \text{od}\}$.

Here for simplicity of intermediate transformations, a call of any type is assumed to be processed only by one free channel (although consideration can be given to models with broadband data calls). Note that the time of channel occupation is determined as a minimum of two random values which describe the duration of call processing (i.e., a time of call processing without considering a handover effect) and a time of a mobile user staying within a cell. Here, as in most of the known papers, for obtaining analytically tractable results, the mentioned two random variables are assumed to be exponentially distributed. Then the time distribution functions of heterogeneous calls occupying the channels are also exponential but with different average values. Assume that the average time of channel occupation for one voice call (a new one or a handover) equals $1/\mu_\text{v}$ and the corresponding metric for data calls (the new ones or handover) equals $1/\mu_\text{d}$.

Now we are passing to describing the call admission control (CAC) scheme. First we note that if upon arrival of od-call there exists at least one free channel of the system, then such a call is taken for service; otherwise it is lost. If upon arrival of hd-call all channels of cell are busy, then it joins a queue (of finite or infinite length).

Access of voice calls is performed by the following scheme:

- If at the moment of ov-call entering the total number of voice calls is less than $R_{ov}, 0 < R_{ov} < N$, then it is taken for service; otherwise it is lost.
- If at the moment of hv-call entering the total number of voice calls is less than $R_{hv}, R_{ov} \leq R_{hv} < N$, then it is taken for service; otherwise it is lost.

The problem consists in finding QoS metrics for the given system, i.e., the blocking (loss) probabilities of calls of each type and the average number of hd-calls in a queue.

In the given system a stationary mode exists if the following condition holds true: $\lambda_d < (N - R_{hv})\mu_d$ where λ_d is the total arrival intensity of data calls.

In a stationary mode a cell state at the arbitrary time instant is described by 2-D vector $\boldsymbol{n} = (n_v, n_d)$, where components n_v and n_d denote the number of voice calls in channels and the total number of data calls in the system, respectively. Since voice calls are served in a block mode and the system is conservative (i.e., with the queue of hd-calls available, channel idling is not allowed), then in any possible state $\boldsymbol{n}$ the number of d-calls in channels (n_d^s) and in the queue (n_d^q) is determined as follows:

$$n_d^s = \min\{N - n_v, n_d\}, n_d^q = (n_v + n_d - N)^+,$$

where $x^+ = \max\{0, x\}$

Hence, the state space of the given 2-D MC is of the form

$$S = \{\boldsymbol{n} : n_v = 0, 1, \ldots, R_{hv}, n_d = 0, 1, \ldots; n_v + n_d^s \leq N\}. \tag{4.1}$$

According to the introduced CAC scheme the nonnegative elements of Q-matrix of the given 2-D MC $q(\boldsymbol{n}, \boldsymbol{n}'), \boldsymbol{n}, \boldsymbol{n}' \in S$ are determined from the following relations:

$$q(\boldsymbol{n}, \boldsymbol{n}') = \begin{cases} \lambda_v & \text{if } n_v \leq R_{ov} - 1, \boldsymbol{n}' = \boldsymbol{n} + \boldsymbol{e}_1, \\ \lambda_{hv} & \text{if } R_{ov} \leq n_v \leq R_{hv} - 1, \boldsymbol{n}' = \boldsymbol{n} + \boldsymbol{e}_1, \\ \lambda_d & \text{if } n_v + n_d^s < N, \boldsymbol{n}' = \boldsymbol{n} + \boldsymbol{e}_2, \\ \lambda_{hv} & \text{if } n_v + n_d^s \geq N, \boldsymbol{n}' = \boldsymbol{n} + \boldsymbol{e}_2, \\ n_v\mu_v & \text{if } \boldsymbol{n}' = \boldsymbol{n} - \boldsymbol{e}_1, \\ n_d^s\mu_d & \text{if } \boldsymbol{n}' = \boldsymbol{n} - \boldsymbol{e}_2, \\ 0 & \text{in other cases,} \end{cases} \tag{4.2}$$

where $\lambda_d = \lambda_{od} + \lambda_{hd}, \lambda_v = \lambda_{ov} + \lambda_{hv}$.

Let P_x be the stationary blocking probability of the calls of type $x, x \in \{\text{hv, ov, hd, od}\}$. Then, in view of the proposed CAC scheme, we obtain that the above-indicated QoS metrics are determined as the corresponding marginal distributions of the 2-D MC. Indeed, hv-calls are lost when the following events occurred: (a) at the moment of hv-call entering the number of v-calls in channels equals R_{hv} and (b) at the moment of hv-call entering all channels are busy regardless of the number of v-calls in the system. Therefore, the loss probability of hv-calls is determined as

$$P_{\mathrm{hv}} = \sum_{\boldsymbol{n}\in S} p(\boldsymbol{n})\big(\delta(n_{\mathrm{v}}, R_{\mathrm{hv}}) + (1-\delta(n_{\mathrm{v}}, R_{\mathrm{hv}}))I(n_{\mathrm{v}} + n_{\mathrm{d}}^{\mathrm{s}} \geq N)\big). \tag{4.3}$$

In formula (4.3) the first term of the sum defines the probability of event (a) occurrence, and the second, the event (b) occurrence.

Using the analogous line of reasoning, we put down the loss probability of ov-calls:

$$P_{\mathrm{ov}} = \sum_{\boldsymbol{n}\in S} p(\boldsymbol{n})\big(I(n_{\mathrm{v}} \geq R_{\mathrm{ov}}) + I(n_{\mathrm{v}} < R_{\mathrm{ov}})I(n_{\mathrm{v}} + n_{\mathrm{d}}^{\mathrm{s}} \geq N)\big). \tag{4.4}$$

Losses of od-calls occur only when at the moment of the given type call entering all channels of cell are busy, i.e.,

$$P_{\mathrm{od}} = \sum_{\boldsymbol{n}\in S} p(\boldsymbol{n})I\big(n_{\mathrm{v}} + n_{\mathrm{d}}^{\mathrm{s}} \geq N\big). \tag{4.5}$$

The average number of hd-calls in the queue (L_{hd}) is determined as follows:

$$L_{\mathrm{hd}} = \sum_{k=1}^{\infty} k\xi(k), \tag{4.6}$$

where $\xi(k) = \sum_{\boldsymbol{n}\in S} p(\boldsymbol{n})\delta(n_{\mathrm{d}}^{\mathrm{q}}, k)$.

Hence, to find the required QoS metrics by the expressions (4.3)–(4.6) one needs to calculate the steady-state probabilities $p(\boldsymbol{n}), \boldsymbol{n} \in S$. It is known that the mentioned probabilities satisfy the corresponding system of global balance equations (SGBE).

Using Kolmogorov's theorem [7] on reversibility of 2-D MC it is easy to show that in the given system the condition of local balance does not hold true. In other words, there is no multiplicative solution to the mentioned SGBE for stationary state probabilities. The alternative approach to solution of the given problem based on the 2-D generating function method is cumbersome and not constructive even for models of small dimension and with the simplest call access schemes. In this connection further we propose another approach which allows one to develop simple calculation procedure for approximate calculation of QoS metrics (4.3)–(4.6).

The given algorithms have high accuracy for the models in which parameters of heterogeneous traffics essentially differ from each other. The last condition almost always holds true in multiservice communication networks since there the average time of voice call transmission is measured by a few minutes, whereas the transmission of packages of data calls takes a few microseconds on average [2]. Moreover, in modern (and the next-generation networks, the ones that are expected) communication networks the data calls form a larger part of a common traffic [3]. In other words, further we take the following assumption: $\lambda_{\mathrm{d}} \gg \lambda_{\mathrm{v}}, \mu_{\mathrm{d}} \gg \mu_{\mathrm{v}}$.

The following presentation implies that the final results do not depend directly on $\lambda_v, \lambda_d, \mu_v$ and μ_d and depend only on their ratios.

Consider the following splitting of state space (4.1):

$$S = \cup_{k=0}^{R_{hv}} S_k, \; S_i \cap S_j = \varnothing, \quad i \neq j, \tag{4.7}$$

where $S_k = \{\boldsymbol{n} \in S : n_v = k\}$. In other words, one performs the partitioning of state space of the model by the value of the first component of 2-D state vector.

Note 4.1 Under fulfillment of the above assumption one preserves the basic principle of applicability of space merging algorithms [14]: in splitting (4.7) the state space of initial model has been partitioned in such classes that the transition probabilities between states inside classes essentially exceed transition probabilities between states from different classes.

The classes of states S_k are combined in merged states $\langle k \rangle$ and in the initial state space S, the following merging function is constructed:

$$U(\boldsymbol{n}) = \langle k \rangle \text{ if } \boldsymbol{n} \in S_k, \quad k = 0, 1, \ldots, R_{hv}. \tag{4.8}$$

The function (4.8) defines the merged model which represents the 1-D BDP with state space $\widetilde{S} = \{\langle k \rangle : k = 0, 1, \ldots, R_{hv}\}$. Hence, steady-state probabilities of initial model are approximately determined as follows (see Appendix of the book [14]):

$$p(k,i) \approx \rho_k(i)\pi(\langle k \rangle), (k,i) \in S_k, \quad k = 0, 1, \ldots, R_{hv}, \quad i = 0, 1, \ldots, \tag{4.9}$$

where $\rho_k(i) : (k,i) \in S_k$ and $\{\pi(\langle k \rangle) : \langle k \rangle \in \widetilde{S}\}$ are the stationary distributions of state probabilities inside the class S_k and the merged model, respectively.

The nonnegative elements of Q-matrix of split model with state space S_k are denoted by $q_k(i,j)$. Considering Eqs. (4.2) and (4.7) we obtain that these parameters are determined from the following relations:

$$q_k(i,j) = \begin{cases} \lambda_d & \text{if } i \leq N-k-1, j = i+1, \\ \lambda_{hd} & \text{if } i \geq N-k, j = i+1, \\ \min(i, N-k)\mu_d & \text{if } j = i-1, \\ 0 & \text{in other cases.} \end{cases} \tag{4.10}$$

From formula (4.10), one can see that the stationary distribution of state probabilities of the split model with the state space S_k coincides with stationary distribution of state probabilities of queueing model $M|M|N-k|\infty$ with state-dependent arriving intensity and constant service intensity of one channel which is equal μ_d. Hence, with ergodicity condition being fulfilled, i.e., at $\lambda_d < (N-k)\mu_d$ stationary state probabilities of the split model with the state space S_k are determined as follows:

$$\rho_k(k) = \begin{cases} \dfrac{\nu_d^i}{i!}\rho_k(0) & \text{if } 1 \le i \le N-k, \\ \left(\frac{\nu_d}{\nu_{hd}}\right)^{N-k} \dfrac{(N-k)^{N-k}}{(N-k)!} \left(\frac{\nu_{hd}}{N-k}\right)^i \rho_k(0) & \text{if } i \ge N-k+1, \end{cases} \tag{4.11}$$

where

$$\nu_d = \lambda_d/\mu_d,\ \nu_{hd} = \lambda_{hd}/\mu_d, \rho_k(0) = \left(\sum_{i=0}^{N-k} \frac{\nu_d^i}{i!} + \frac{\nu_d^{N-k}}{(N-k)!} \frac{\nu_{hd}}{N-k-\nu_{hd}}\right)^{-1}.$$

Since the ergodicity condition $\nu_d < N-k$ must to be hold true for each $k = 0, 1, \ldots, R_{hv}$, then we obtain the ergodicity condition of the initial model $\nu_d < N - R_{hv}$ (above this condition was determined by intuition). Then under fulfillment of this condition taking into account Eq. (4.11) from Eq. (4.2), we obtain the following relations for calculating nonnegative elements of Q-matrix for the merged model (for brevity of presentation here and further the corresponding mathematical transformations are omitted):

$$q(\langle k \rangle, \langle k' \rangle) = \begin{cases} \lambda_v \alpha_k & \text{if } 0 \le k \le R_{ov} - 1,\ k' = k+1, \\ \lambda_{hv} \alpha_k & \text{if } R_{ov} \le k \le R_{hv} - 1,\ k' = k+1, \\ k\mu_v & \text{if } k' = k-1, \\ 0 & \text{in other cases,} \end{cases} \tag{4.12}$$

where $\alpha_k = \rho_k(0) \sum_{i=0}^{N-k-1} \nu_d^i / i!\,,\quad k = 0, 1, \ldots, R_{hv} - 1.$

The relations (4.12) allow one to determine steady-state probabilities of merged model which is described by 1-D BDP. Hence, the required distribution of state probabilities of the merged model is determined in the following way:

$$\pi(\langle k \rangle) = \begin{cases} \dfrac{\nu_v^k}{k!} \prod_{i=0}^{k-1} \alpha_i \pi(\langle 0 \rangle) & \text{if } 1 \le k \le R_{ov}, \\ \left(\frac{\nu_v}{\nu_{hv}}\right)^{R_{ov}} \dfrac{\nu_{hv}^k}{k!} \prod_{i=0}^{k-1} \alpha_i \pi(\langle 0 \rangle) & \text{if } R_{ov} + 1 \le k \le R_{hv}, \end{cases} \tag{4.13}$$

where $\nu_v = \lambda_v/\mu_v$, $\nu_{hv} = \lambda_{hv}/\mu_v$ and $\pi(\langle 0 \rangle)$ is determined from normalizing condition, i.e., $\sum_{k=0}^{R_{hv}} \pi(\langle k \rangle) = 1$.

Here and further we put $\prod_{i=m}^{n} x_i = 1$ if $n < m$.

Using Eqs. (4.11)–(4.13) and after certain mathematical transformations we finally obtain the following approximate formulas for calculating QoS metrics from expressions (4.3)–(4.6):

$$P_{\mathrm{hv}} = \sum_{n \in S_{R_{\mathrm{hv}}}} p(n) + \sum_{k=0}^{R_{\mathrm{hv}}-1} \sum_{n \in S_k} p(n) I(n_{\mathrm{v}} + n_{\mathrm{d}}^{\mathrm{s}} \geq N) \approx \pi(\langle R_{\mathrm{hv}} \rangle) + \sum_{k=0}^{R_{\mathrm{hv}}-1} \pi(\langle k \rangle) \left(1 - \rho_k(0) \sum_{i=0}^{N-k-1} \frac{v_{\mathrm{d}}^i}{i!} \right); \tag{4.14}$$

$$P_{\mathrm{ov}} = \sum_{k=R_{\mathrm{ov}}}^{R_{\mathrm{hv}}} \sum_{n \in S_k} p(n) + \sum_{k=0}^{R_{\mathrm{ov}}-1} \sum_{n \in S_k} p(n) I(n_{\mathrm{v}} + n_{\mathrm{d}}^{\mathrm{s}} \geq N) \approx \sum_{k=R_{\mathrm{ov}}}^{R_{\mathrm{hv}}} \pi(\langle k \rangle) + \sum_{k=0}^{R_{\mathrm{ov}}-1} \pi(\langle k \rangle) \left(1 - \rho_k(0) \sum_{i=0}^{N-k-1} \frac{v_{\mathrm{d}}^i}{i!} \right); \tag{4.15}$$

$$P_{\mathrm{od}} = \sum_{k=0}^{R_{\mathrm{hv}}} \sum_{n \in S_k} p(n) I(n_{\mathrm{v}} + n_{\mathrm{d}}^{\mathrm{s}} \geq N) \approx \sum_{k=0}^{R_{\mathrm{hv}}} \pi(\langle k \rangle) \left(1 - \rho_k(0) \sum_{i=0}^{N-k-1} \frac{v_{\mathrm{d}}^i}{i!} \right); \tag{4.16}$$

$$L_{\mathrm{hd}} = \sum_{k=1}^{\infty} k \sum_{n \in S} p(n) \delta(n_{\mathrm{d}}^{\mathrm{q}}, k) \approx \sum_{k=1}^{\infty} k \sum_{i=0}^{R_{\mathrm{hv}}} \rho_i (N + k - i) \pi(\langle i \rangle) = \sum_{k=0}^{R_{\mathrm{hv}}} \pi(\langle k \rangle) \sum_{i=1}^{\infty} i \rho_k (N - k + i) = \sum_{k=0}^{R_{\mathrm{hv}}} \pi(\langle k \rangle) L(M/M/N - k/\infty). \tag{4.17}$$

In the last formula $L(M|M|N - k|\infty)$ represents the average length of queue in the above-described model $M|M|N - k|\infty$ (see the description of split model with state space S_k), i.e.,

$$L(M/M/N - k/\infty) = \rho_k(0) \frac{v_{\mathrm{d}}^{N-k}}{(N-k)!} \frac{v_{\mathrm{hd}}(k)}{(1 - v_{\mathrm{hd}}(k))^2},$$

where $v_{\mathrm{hd}}(k) = v_{\mathrm{hd}}/(N - k)$.

Now consider a special case of the studied CAC scheme in which there was no difference between new and handover calls of voice traffic, i.e., it is assumed that $R_{\mathrm{ov}} = R_{\mathrm{hv}}$. In other words, the proposed call access scheme contains only one threshold parameter. The stationary distribution of state probabilities of split model in this case is also determined by the relations (4.11). However, state probabilities of merged model in the given case are determined as

$$\pi(\langle k \rangle) = \frac{v_{\mathrm{v}}^k}{k!} \prod_{i=0}^{k-1} \alpha_i \pi(\langle 0 \rangle), \quad k = 1, \ldots, R_{\mathrm{hv}},$$

where $\pi(\langle 0 \rangle) = \left(\sum_{k=0}^{R_{\mathrm{hv}}} (v_{\mathrm{v}}^k / k!) \prod_{i=1}^{k-1} \alpha_i \right)^{-1}$.

For this case from Eqs. (4.14) and (4.15), we have

$$P_{\text{ov}} = P_{\text{hv}} \approx \pi(\langle R_{\text{hv}} \rangle) + \sum_{k=0}^{R_{\text{hv}}-1} \pi(\langle k \rangle)\left(1 - \rho_k(0) \sum_{i=0}^{N-k-1} \frac{v_{\text{d}}^i}{i!}\right).$$

The values of P_{od} and L_{hd} are determined from Eqs. (4.16) and (4.17), respectively.

The proposed approach can be also applied in models with limited queue of hd-calls. Let the maximal buffer size for hd-calls waiting be equal to R_{hd}. Then by the above procedure we obtain that in the given model the stationary distribution of state probabilities of split model with the state space S_k coincides with the stationary probability distribution of states for the queueing model $M|M|N - k|R_{\text{hd}}$ with state-dependent arrival rate and the constant intensity of one channel service equal μ_{d}. In models with a limited queue, the condition of stationary mode always holds true and, hence, steady-state probabilities of split model with state space S_k are determined as

$$\rho_k(k) = \begin{cases} \frac{v_{\text{d}}^i}{i!}\rho_k(0) & \text{if } 1 \le i \le N-k, \\ \left(\frac{v_{\text{d}}}{v_{\text{hd}}}\right)^{N-k} \frac{(N-k)^{N-k}}{(N-k)!}\left(\frac{v_{\text{hd}}}{N-k}\right)^i \rho_k(0) & \text{if } N-k+1 \le i \le N-k+R_{\text{hd}}, \end{cases} \tag{4.18}$$

where $\rho_k(0)$ is determined from normalizing condition, i.e., $\sum_{i=0}^{N-k+R_{\text{hd}}} \rho_k(i) = 1$.

Further by Eq. (4.13) we determine the stationary distribution of merged model taking into account that the parameters α_k are calculated in view of Eq. (4.18). The QoS metrics (4.3)–(4.5) are also determined by the formulas (4.14)–(4.16), respectively. In the given model the average length of queue of hd-calls is calculated as follows:

$$L_{\text{hd}} \approx \sum_{k=0}^{R_{\text{hv}}} \pi(\langle k \rangle) L(M/M/N-k/R_{\text{hd}}), \tag{4.19}$$

where $L(M|M|N - k|R_{\text{hd}})$ is the average length of queue in the above model $M|M|N - k|R_{\text{hd}}$, i.e.,

$$L(M/M/N-k/R_{\text{hd}}) = \rho_k(0)\frac{v_{\text{d}}^{N-k}}{(N-k)!}\sum_{i=1}^{N+R_{\text{hd}}-k} i(v_{\text{hd}}(k))^i.$$

Note that in the given model a new QoS metric which denotes the blocking probability of handover data calls appears. It is determined as follows:

$$P_{\mathrm{hd}} \approx \sum_{k=0}^{R_{\mathrm{hv}}} \rho_k (N - k + R_{\mathrm{hd}})\pi(\langle k \rangle). \tag{4.20}$$

The model with a limited queue of impatient hd-calls can be studied in analogous way, and in this case one also manages to obtain the explicit formulas. Regardless of the fact that for the model with unlimited queue of impatient hd-calls one is not able to obtain explicit formulas, in the given case it is possible to use the known approximate scheme ([14], Chap. 2]).

Now consider the second scheme to restrict the access of voice calls to channels in which the decision on access of heterogeneous v-calls is based on the total number of busy channels of the cell (guard channel scheme). First of all note that the access scheme for data calls remains same as in previous scheme.

The access of voice calls in this scheme is performed by the following rules:

- If at the moment of arrival of ov-call, the total number of busy channels is less than $G_{\mathrm{ov}}, 0 < G_{\mathrm{ov}} < N$, then it is accepted for service; otherwise it is rejected.
- If at the moment of arrival of hv-call, the total number of busy channels is less than $G_{\mathrm{hv}}, G_{\mathrm{ov}} \le G_{\mathrm{hv}} < N$, then it is accepted for service; otherwise it is rejected.

In the given scheme the stationary mode in the cell exists under fulfillment of condition $\lambda_{\mathrm{d}} < (N - G_{\mathrm{hv}})\mu_{\mathrm{d}}$, where λ_{d} is the total intensity of data calls. Then in a stationary mode the cell state at the arbitrary time instant is described by two-dimensional vector $\boldsymbol{n} = (n_{\mathrm{v}}, n_{\mathrm{d}})$, where n_{v} and n_{d} indicate the number of voice calls in channels and the total number of data calls in the cell, respectively. Hence, the space of states of the given 2-D MC is determined similar to Eq. (4.1), i.e.,

$$S = \left\{ \boldsymbol{n} : n_{\mathrm{v}} = 0, 1, \ldots, G_{\mathrm{hv}}, n_{\mathrm{d}} = 0, 1, \ldots; n_{\mathrm{v}} + n_{\mathrm{d}}^{\mathrm{s}} \le N \right\}.$$

According to the given access scheme the nonnegative elements of Q-matrix of the given 2-D Markov chain $q(\boldsymbol{n}, \boldsymbol{n}')$, $\boldsymbol{n}, \boldsymbol{n}' \in S$ are determined from the following relations:

$$q(\boldsymbol{n}, \boldsymbol{n}') = \begin{cases} \lambda_{\mathrm{v}} & \text{if}\, n_{\mathrm{v}} + n_{\mathrm{d}}^{\mathrm{s}} \le G_{\mathrm{ov}} - 1,\, \boldsymbol{n}' = \boldsymbol{n} + \boldsymbol{e}_1, \\ \lambda_{\mathrm{hv}} & \text{if}\, G_{\mathrm{ov}} \le n_{\mathrm{v}} + n_{\mathrm{d}}^{\mathrm{s}} \le G_{\mathrm{hv}} - 1,\, \boldsymbol{n}' = \boldsymbol{n} + \boldsymbol{e}_1, \\ \lambda_{\mathrm{d}} & \text{if}\, n_{\mathrm{v}} + n_{\mathrm{d}}^{\mathrm{s}} < N,\, \boldsymbol{n}' = \boldsymbol{n} + \boldsymbol{e}_2, \\ \lambda_{\mathrm{hv}} & \text{if}\, n_{\mathrm{v}} + n_{\mathrm{d}}^{\mathrm{s}} \ge N,\, \boldsymbol{n}' = \boldsymbol{n} + \boldsymbol{e}_2, \\ \lambda_{\mathrm{v}}\mu_{\mathrm{v}} & \text{if}\, \boldsymbol{n}' = \boldsymbol{n} - \boldsymbol{e}_1, \\ n_{\mathrm{d}}^{\mathrm{s}}\mu_{\mathrm{d}} & \text{if}\, \boldsymbol{n}' = \boldsymbol{n} - \boldsymbol{e}_2, \\ 0 & \text{in other cases.} \end{cases} \tag{4.21}$$

Then, according to the proposed call access scheme, we find that the loss probabilities of heterogeneous v-calls are determined as the corresponding marginal distributions of initial 2-D MC, i.e.,

$$P_{\text{hv}} = \sum_{\boldsymbol{n} \in S} p(\boldsymbol{n}) I\left(n_{\text{v}} + n_{\text{d}}^{\text{s}} \geq G_{\text{hv}}\right) \tag{4.22}$$

$$P_{\text{ov}} = \sum_{\boldsymbol{n} \in S} p(\boldsymbol{n}) I\left(n_{\text{v}} + n_{\text{d}}^{\text{s}} \geq G_{\text{ov}}\right) \tag{4.23}$$

The remains QoS metrics, i.e., the loss probability of od-calls and the average number of hd-calls in the queue (L_{hd}), are determined by Eqs. (4.5) and (4.6), respectively.

As it was mentioned above, the approach to finding the state probabilities based on application of two-dimensional generating function method is cumbersome and ineffective even for simple access schemes. In this connection further we use the above-developed approach, which allows one to develop simple algorithms for approximate calculation of QoS metrics for the given call access scheme.

Under the above-accepted assumption related to ratios of loading parameters, consider the following splitting of state space:

$$S = \cup_{k=0}^{G_{\text{hv}}} S_k, \; S_i \cap S_j = \varnothing, \quad i \neq j, \tag{4.24}$$

where $S_k = \{\boldsymbol{n} \in S : n_{\text{v}} = k\}$.

As the scheme of splitting of the state space completely defines the split and merged models, some intermediate stages are omitted below.

Hence, with the ergodicity condition fulfilled (i.e., at $\lambda_{\text{d}} < (N-k)\mu_{\text{d}}$), the steady-state probabilities of the split model with the state space S_k, $k = 0, 1, \ldots, G_{\text{hv}}$ are determined by the similar formulas (4.11). Since the ergodicity condition $v_{\text{d}} < N - k$ should be fulfilled for each $k = 0, 1, \ldots, G_{\text{hv}}$ then the ergodicity condition for the initial model is obtained: $v_{\text{d}} < N - G_{\text{hv}}$. Then with this condition fulfilled and in view of Eq. (4.11) from Eq. (4.21), we obtain the following relations to calculate the nonnegative elements of Q-matrix of the merged model:

$$q(\langle k \rangle, \langle k' \rangle) = \begin{cases} \lambda_{\text{v}} \beta_k & \text{if } 0 \leq k \leq G_{\text{ov}} - 1, \, k' = k+1, \\ \lambda_{\text{hv}} \beta_k & \text{if } G_{\text{ov}} \leq k \leq G_{\text{hv}} - 1, \, k' = k+1, \\ k\mu_{\text{v}} & \text{if } k' = k-1, \\ 0 & \text{in other cases,} \end{cases}$$

where

$$\beta_k = \begin{cases} \rho_k(0) \displaystyle\sum_{i=0}^{G_{\text{ov}}-k-1} \frac{v_{\text{d}}^i}{i!} & \text{if } 0 \leq k \leq G_{\text{ov}} - 1, \\ \rho_k(0) \displaystyle\sum_{i=0}^{G_{\text{hv}}-k-1} \frac{v_{\text{d}}^i}{i!} & \text{if } G_{\text{ov}} \leq k \leq G_{\text{hv}} - 1. \end{cases}$$

These relations allow one to determine the steady-state probabilities of the merged model, which is described by 1-D BDP, i.e.,

$$\pi(\langle k\rangle)=\begin{cases}\frac{\nu_{\mathrm{v}}^{k}}{k!}\prod_{i=0}^{k-1}\beta_i\pi(\langle 0\rangle) & \text{if } 1\le k\le G_{\mathrm{ov}},\\ \left(\frac{\nu_{\mathrm{v}}}{\nu_{\mathrm{hv}}}\right)^{G_{\mathrm{ov}}}\frac{\nu_{\mathrm{hv}}^{k}}{k!}\prod_{i=0}^{k-1}\beta_i\pi(\langle 0\rangle) & \text{if } G_{\mathrm{ov}}+1\le k\le G_{\mathrm{hv}},\end{cases} \tag{4.25}$$

where $\pi(\langle 0\rangle)$ is determined from normalizing condition, i.e., $\sum_{k=0}^{G_{\mathrm{hv}}}\pi(\langle k\rangle)=1$.

After certain transformations we obtain the following approximate formulas for calculating QoS metrics of the given CAC scheme:

$$P_{\mathrm{hv}}\approx\pi(\langle G_{\mathrm{hv}}\rangle)+\sum_{k=0}^{G_{\mathrm{hv}}-1}\pi(\langle k\rangle)\left(1-\rho_k(0)\sum_{i=0}^{G_{\mathrm{hv}}-k-1}\frac{\nu_{\mathrm{d}}^{i}}{i!}\right); \tag{4.26}$$

$$P_{\mathrm{ov}}\approx\sum_{k=G_{\mathrm{ov}}}^{G_{\mathrm{hv}}}\pi(\langle k\rangle)+\sum_{k=0}^{G_{\mathrm{ov}}-1}\pi(\langle k\rangle)\left(1-\rho_k(0)\sum_{i=0}^{G_{\mathrm{ov}}-k-1}\frac{\nu_{\mathrm{d}}^{i}}{i!}\right); \tag{4.27}$$

$$P_{\mathrm{od}}\approx\sum_{k=0}^{G_{\mathrm{hv}}}\pi(\langle k\rangle)\left(1-\rho_k(0)\sum_{i=0}^{N-k-1}\frac{\nu_{\mathrm{d}}^{i}}{i!}\right); \tag{4.28}$$

$$L_{\mathrm{hd}}\approx\sum_{k=0}^{G_{\mathrm{hv}}}\pi(\langle k\rangle)\rho_k(0)\frac{\nu_{\mathrm{d}}^{N-k}}{(N-k)!}\frac{\nu_{\mathrm{hd}}(k)}{(1-\nu_{\mathrm{hd}}(k))^{2}}. \tag{4.29}$$

Here we can also consider some special cases of the CAC scheme under study. For instance, when there is no differences between new and handover calls of voice traffic (i.e., $G_{\mathrm{ov}}=G_{\mathrm{hv}}$ and $G_{\mathrm{hv}}<N$), the steady-state probabilities of the merged model are determined as follows:

$$\pi(\langle k\rangle)=\frac{\nu_{\mathrm{v}}^{k}}{k!}\prod_{i=0}^{k-1}\beta_i\pi(\langle 0\rangle),\quad k=1,\ldots,G_{\mathrm{hv}},$$

where

$$\beta_i=\rho_i(0)\sum_{j=0}^{G_{\mathrm{hv}}-i-1}\frac{\nu_{\mathrm{d}}^{j}}{j!}.$$

For this case from Eqs. (4.26) and (4.27), we have

$$P_{\mathrm{ov}}=P_{\mathrm{hv}}\approx\pi(\langle G_{\mathrm{hv}}\rangle)+\sum_{k=0}^{G_{\mathrm{hv}}-1}\pi(\langle k\rangle)\left(1-\rho_k(0)\sum_{i=0}^{G_{\mathrm{hv}}-k-1}\frac{\nu_{\mathrm{d}}^{i}}{i!}\right).$$

The values of P_{od} and L_{hd} are determined from Eqs. (4.28) and (4.29), respectively.

The proposed approach can be applied also to calculating QoS metrics of the models with the given CAC scheme and with the limited queue of hd-calls. Since the analogous formulas for calculating the models with the cutoff strategy have been presented above, they are omitted here.

4.1.2 Selection of Effective Values of Access Scheme Parameters

The formulas obtained in Sect. 4.1.1 allow one to study the behavior of QoS metrics for fixed values of number of channels and parameters of the introduced access schemes. However, of certain scientific and practical interest are the problems satisfying the given level of QoS for heterogeneous calls due to choice of appropriate values of CAC being used.

There can be different statements of problems of determining such values of threshold parameters. In what follows consideration is given to problems of searching for the set of values of parameters of the introduced CAC schemes satisfying the given level of service quality of heterogeneous calls. This set (if it is not empty) will be called the set of efficient values (SEVs) of access scheme parameters.

First consider CAC based on cutoff scheme. Verbal definition of the problem considered consists in the following. Let under the fixed loads there be given upper limits for possible values of loss probabilities of heterogeneous calls and the average length of queue of hd-calls (at the same time the last limitation means that there is given the limitation on the average waiting time of hd-calls in a queue). One needs to find such values of parameters R_{ov} and R_{hv} that the given limitations would be satisfied.

Note that for small values of N this problem solution can be found by simple enumeration of all possible combinations of the values of parameters R_{ov} and R_{hv}. However, with N growing such approach is not efficient and sometimes not feasible. Hence, further we propose the algorithmic approach to solution of the stated problem which does not use the complete enumeration of variants.

Mathematically the mentioned problem is written as follows: one needs to find the pairs (R_{ov}, R_{hv}), $R_{ov} \le R_{hv}$, satisfying the following limitations:

$$P_x \le \varepsilon_x, \quad x \in \{\text{ov}, \text{hv}, \text{od}\}, \tag{4.30}$$

$$L_{hd} \le l_{hd}, \tag{4.31}$$

where $\varepsilon_{hv}, \varepsilon_{ov}, \varepsilon_{od}, l_{hd}$ are the given values.

One of the possible algorithms of solving the problem (4.30), (4.31) based on application of monotony property of studied QoS metrics is described further. The main idea of this iterative algorithm consists in the fact that for each fixed value of parameter R_{hv} the search for SEV is performed due to the choice of corresponding values of parameter R_{ov}. In other words, the functions involved in the given problem have just one argument R_{ov}. For simplicity of presentation this argument is explicitly indicated in records of these functions.

For generality we consider the kth iteration, $k=1,2,\ldots,R_{hv}$:

Step 1. We put $R_{hv}=k$ and verify the condition $P_{ov}(k)\le\varepsilon_{ov}$. If it holds true, then we pass to the next step; otherwise, for the given value of R_{hv} the problem has no solution.

Step 2. In parallel the following problems are solved:

$$R_{ov}^{*}=\arg\min_{R_{ov}\in[1,k]}\{P_{ov}(R_{ov})\le\varepsilon_{ov}\}; \tag{4.32}$$

$$R_{1}=\arg\min_{R_{ov}\in[1,k]}\{P_{hv}(R_{ov})\le\varepsilon_{hv}\}; \tag{4.33}$$

$$R_{2}=\arg\max_{R_{ov}\in[1,k]}\{P_{od}(R_{ov})\le\varepsilon_{od}\}; \tag{4.34}$$

$$R_{3}=\arg\max_{R_{ov}\in[1,k]}\{L_{hd}(R_{ov})\le l_{od}\}. \tag{4.35}$$

We put $R_{ov}^{**}:=\min\{R_1,R_2,R_3\}$.

Step 3. The required interval of appropriate values of R_{ov} for the given value of R_{hv} is determined as $[1,R_{ov}^{**}]\cap[R_{ov}^{*},k]$.

Step 4. If $R_{hv}<N$, then we put $R_{hv}:=R_{hv}+1$ and pass to the step 1. Otherwise the algorithm operation terminates.

Note 4.2 On the basis of monotony property of functions studied to solve the problems (4.32)–(4.35), one can employ a dichotomy method.

Therefore, for every fixed value of threshold R_{hv}, the algorithm given above searches for the set of admissible values of R_{ov} (if they exist) so that by integrating all obtained solutions to obtain SEV of threshold parameters.

With application of developed algorithm the numerical experiments were performed. The initial data of test problems (4.30), (4.31) for the hypothesized model is chosen in the following way: $N=20$; $\lambda_{ov}=0.2$; $\lambda_{hv}=0.1$; $\lambda_{od}=2$; $\lambda_{hd}=1$; $\mu_{v}=0.1$; $\mu_{d}=2$. For clearness the results of solving the problems (4.30), (4.31) under different limitations on values of loss probability of heterogeneous calls are gathered in Table 4.1, where ∅ denotes that the problem has no solution.

Now consider the similar problem of finding the set of effective values of threshold parameters for CAC based on guard channel scheme. In other words, it is necessary to find the pairs (G_{ov},G_{hv}), $G_{ov}\le G_{hv}$ such that the restrictions (4.30), (4.31) are fulfilled.

To solve this problem one can use the modification of the algorithm developed above. With its help the numerical experiments have been performed. Initial data for test problems for the hypothesize model were selected as above. The results of problem solution under different limitations on values of loss probabilities of heterogeneous calls are demonstrated in Table 4.2.

The behavior of QoS metrics of model with respect to change of variable parameters G_{ov},G_{hv} (in the CAC based on guard channel scheme) and R_{ov},R_{hv} (in the CAC based on cutoff scheme) is identical. Hence, the comparison of SEV

Table 4.1 Solution results for problems (4.30) and (4.31) in CAC based on cutoff scheme

ε_{ov}	ε_{od}	ε_{hv}	ε_{hd}	$[R^*_{ov}, R^{**}_{ov}]$ $R_{hv}=10$	$R_{hv}=13$	$R_{hv}=16$
E-02	4E-07	5E-05	5E-05	Ø	[8, 13]	[8, 16]
E-02	E-07	E-04	E-08	[8, 8]	[8, 13]	[8, 16]
E-02	E-07	E-04	E-10	[8, 8]	[8, 8]	[8, 8]
E-02	E-07	E-04	E-09	[8, 8]	[8, 11]	[8, 11]
E-01	E-06	E-04	E-10	[6, 8]	[6, 8]	[6, 8]
5E-02	4E-09	5E-04	2E-10	[7, 9]	[7, 8]	[7, 8]
8E-02	4E-08	5E-09	5E-10	Ø	Ø	[6, 9]
8E-02	4E-08	5E-05	5E-08	[6, 7]	[6, 13]	[6, 16]
15E-02	4E-07	5E-05	5E-12	[5, 5]	[5, 5]	[5, 5]
15E-02	4E-07	5E-06	5E-11	[5, 5]	[5, 7]	[5, 7]

Table 4.2 Solution results for problems (4.30) and (4.31) in CAC based on guard channel scheme

ε_{ov}	ε_{od}	ε_{hv}	ε_{hd}	$[G^*_{ov}, G^{**}_{ov}]$ $G_{hv}=10$	$G_{hv}=13$	$G_{hv}=16$
15E-02	E-09	E-03	E-11	Ø	[7, 7]	[7, 7]
15E-03	E-08	E-03	E-10	Ø	[10, 11]	[10, 11]
5E-02	E-08	14E-03	E-10	[9, 10]	[9, 10]	[9, 10]
5E-02	5E-07	14E-03	2E-05	[9, 10]	[9, 13]	[9, 16]
5E-02	5E-09	14E-03	2E-05	Ø	[9, 11]	[9, 11]
7E-02	2E-09	4E-03	2E-10	Ø	[8, 10]	[8, 10]
E-02	2E-08	3E-03	2E-08	Ø	[11, 13]	[11, 13]
4E-02	2E-08	12E-03	2E-08	[9, 9]	[9, 13]	[9, 13]
3E-02	2E-08	E-01	2E-08	[10, 10]	[10, 13]	[10, 13]
9E-02	2E-08	E-02	2E-08	[8, 8]	[8, 13]	[8, 13]

under different access schemes when initial data of model remains unchangeable represents practical interest. In this connection, it is worth noting that under all limitations on QoS metrics which are demonstrated in Table 4.1, SEV of the problem for CAC based on guard channels represents the empty set. In other words, it should be expected that for the same values of number of cell channels, loads, and the required change of value ranges of QoS metrics, these limitations will be satisfied by one strategy and will not by the other. Since both strategies have the same degree of implementation complexity, then in each particular case one should make a serious study before selecting the appropriate access scheme.

Note that in practice the loads of heterogeneous traffics change in time. However, the above numerical experiments were performed under fixed loads. Hence, problems of studying sensitivity of efficient values of threshold parameters versus load change are urgency also. In this connection, it is worth noting that the analytical study of the given problem is not feasible in principle. Hence, it can be studied only by application of numerical experiments. Fortunately, the simplicity of proposed numerical procedures to calculating the QoS metrics allows one to solve this problem easily. The experiments performed showed that the efficient values of threshold parameters in both CAC are invariable in sufficiently wide range of load change. It is accounted for by a smooth change of studied QoS metrics with respect to loads of heterogeneous traffics (see next subsection).

4.1.3 Numerical Results

Let us first examine the results of numerical experiments for the CAC based on cutoff scheme. Some results for hypothetical models are depicted in Figs. 4.1, 4.2, and 4.3. They illustrate the plots of the studied QoS metrics for the following initial data: $N=20$; $R_{hv}=14$; $\lambda_{ov}=0.2$; $\lambda_{hv}=0.1$; $\lambda_{od}=5$; $\lambda_{hd}=2$; $\mu_v=0.1$; $\mu_d=0.6$.

In the model considered for the fixed value of the total number of channels (N), one can change the values of two parameters of the given CAC (R_{ov} and R_{hv}). In other words, there exist two degrees of freedom for the given model. Note that the increase of value of one parameter (in admissible domain) favorably affects only the loss probability of calls of the corresponding type.

Hence, in these experiments the increase of value of threshold R_{ov} leads to the decrease of loss probability of ov-calls, with three other QoS metrics (i.e., P_{hv}, P_{od}, and L_{hd}) growing. However, the speeds of their change are different. From Fig. 4.1 one can see that the value of function P_{ov} decreases with high speed especially at values of R_{ov} close to maximal possible value (i.e., R_{hv}), and at the point $R_{ov}=R_{hv}$, the values of functions P_{ov} and P_{hv} are equal (as it was to be expected; see formulas (4.3) and (4.4)). However, speeds of other functions growing are sufficiently low, and for values of R_{ov} close to maximal possible value, they almost do not change. Note that the increase of loads of traffics worsens all QoS metrics (see Figs. 4.2 and 4.3).

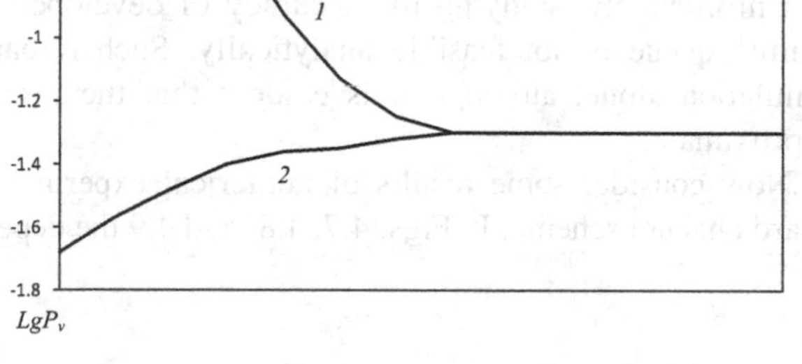

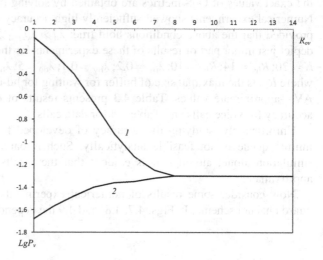

Fig. 4.1 Blocking probability of v-calls versus R_{ov}; 1—P_{ov}, 2—P_{hv}

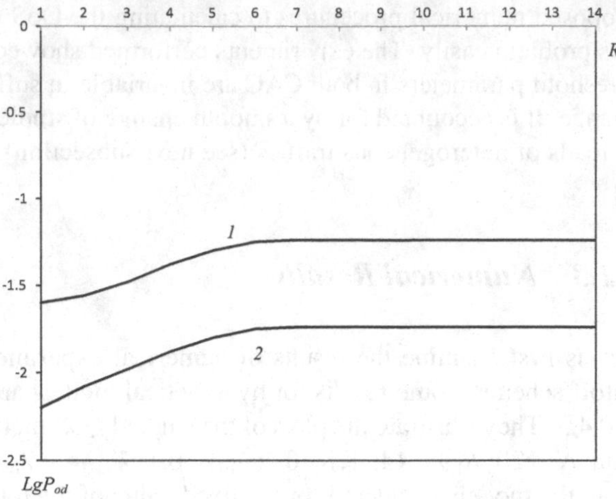

Fig. 4.2 Blocking probability of od-calls versus R_{ov}; 1—$\mu_d = 0.6$, 2—$\mu_d = 0.7$

The behavior of QoS metrics (4.3)–(4.6) with respect to change of traffic is shown in Figs. 4.4, 4.5, and 4.6, where the value to change is the load of voice calls. Here the initial data was selected as follows: $N = 20$; $R_{hv} = 14$; $R_{ov} = 5$; $\lambda_{ov} = 0.2$; $\lambda_{hv} = 0.1$; $\lambda_{od} = 5$; $\lambda_{hd} = 2$; $\mu_d = 2$. This plot analysis shows a smooth growth of functions under study with respect to the increased load of voice calls. The analogous results were obtained with respect to the changed load of data calls. As it was mentioned above (see Sect. 4.1.2), this research is very important in terms of determining sensitivity of optimal (in a certain sense) values of parameters of the applied CAC scheme under the change of traffic parameters.

For models with the finite queue of hd-calls, one can study the accuracy of proposed approximate algorithms of calculating the values of QoS metrics. However, this research is feasible only for small dimensions of state space (4.1) since in this case the exact values of QoS metrics are obtained by solving the corresponding SGBE. Numerical experiments showed sufficiently high accuracy of developed algorithms provided that the above conditions hold true: $\lambda_d \gg \lambda_v, \mu_d \gg \mu_v$. Tables 4.3 and 4.4 depict just minor part of results of these experiments for the model with parameters $N = 20$; $R_{hv} = 14$; $R_{hd} = 10$; $\lambda_{ov} = 0.2$; $\lambda_{hv} = 0.1$; $\lambda_{od} = 5$; $\lambda_{hd} = 2$; $\mu_v = 0.1$; $\mu_d = 0.6$, where R_{hd} is the maximal size of buffer for waiting for hd-calls: EV—exact values; AV—approximate values. Table 4.3 presents results of comparative analysis of accuracy for voice calls and Table 4.4 for data calls.

Unfortunately, studying the accuracy of developed formulas for models with infinite queue is not feasible analytically. Such research can be conducted via simulation alone, although it is evident that the results of simulation are also approximate.

Now consider some results of numerical experiments for the CAC based on guard channel scheme. In Figs. 4.7, 4.8, and 4.9 the dependences of QoS metrics of

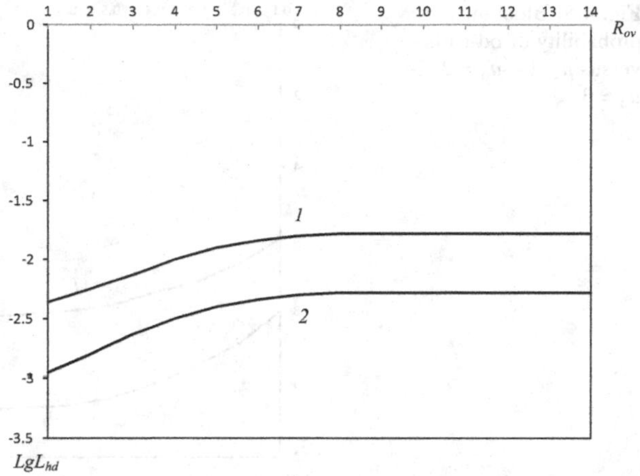

Fig. 4.3 Average length of queue of hd-calls versus R_{ov}; 1—$\mu_d = 0.6$, 2—$\mu_d = 0.7$

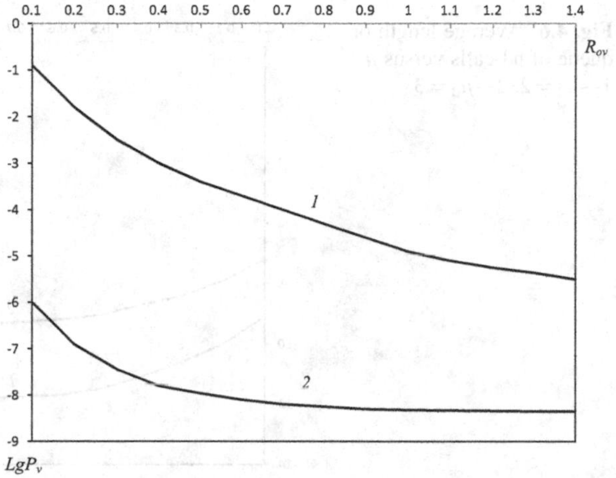

Fig. 4.4 Blocking probability of v-calls versus μ_v; 1—P_{ov}, 2—P_{hv}

the model on the value of two threshold parameters at the fixed value of total number of channels are demonstrated.

Here we present the plots of QoS metrics under study for the following initial data: $N = 20, G_{hv} = 14, \lambda_{ov} = 0.2, \lambda_{hv} = 0.1, \lambda_{od} = 5, \lambda_{hd} = 2, \mu_v = 0.1$. As in case of CAC based on cutoff scheme, the increased value of one of thresholds (in the admissible region) favorably affects only the loss probability of calls of the corresponding type. In these experiments with the increased value of threshold G_{ov} the loss probability of ov-calls decreases; therewith the other three QoS metrics (P_{hv}, P_{od}, L_{hd}) grow. As expected, $P_{hv} = P_{ov}$ when $G_{ov} = G_{hv}$. We also note that the increased value of each traffic intensity and (or) decreased intensity of their service lead to the growing values of the above QoS metrics (see Figs. 4.8 and 4.9).

Results related to the behavior of QoS metrics with respect to change of loads of voice calls are depicted in Figs. 4.10, 4.11, and 4.12, where the initial data are

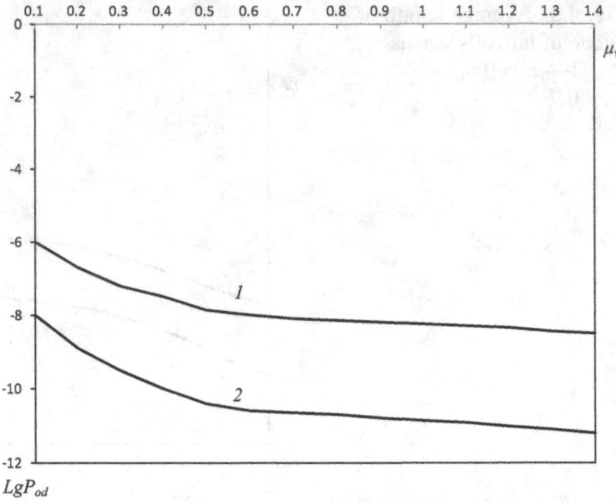

Fig. 4.5 Blocking probability of od-calls versus μ_v; 1—$\mu_d = 2$, 2—$\mu_d = 3$

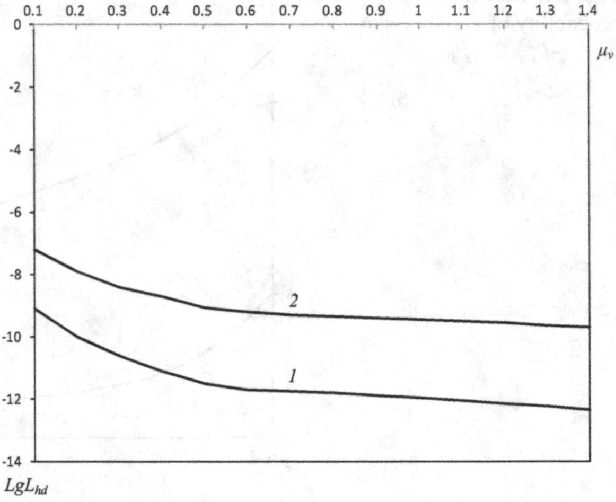

Fig. 4.6 Average length of queue of hd-calls versus μ_v; 1—$\mu_d = 2$, 2—$\mu_d = 3$

chosen as follows: $N = 20, G_{hv} = 14, G_{ov} = 5, \lambda_{ov} = 0.2, \lambda_{hv} = 0.1, \lambda_{od} = 5, \lambda_{hd} = 2$; therewith, in Fig. 4.10 we put $\mu_d = 2$. This plot analysis shows that the QoS metrics under study change smoothly with respect to change of loads of entering traffics (analogous studies have been performed also for data calls loads; however, due to being short the appropriate results are not presented here).

Table 4.3 Comparison for v-calls in CAC based on cutoff scheme

R_{ov}	R_{ov} EV	R_{ov} AV	R_{hv} EV	R_{hv} AV
1	0.851256	0.844055	0.008129	0.007523
2	0.628789	0.624688	0.009611	0.009503
3	0.422075	0.417093	0.012315	0.011958
4	0.273341	0.252182	0.014603	0.014415
5	0.140492	0.139387	0.017012	0.016435
6	0.078063	0.073291	0.017309	0.017787
7	0.043031	0.040325	0.018742	0.018523
8	0.029452	0.026328	0.018901	0.018853
9	0.023543	0.021232	0.018952	0.018979
10	0.021788	0.019624	0.019431	0.019023
11	0.021631	0.019179	0.019479	0.019037
12	0.021607	0.019071	0.019481	0.019042
13	0.021598	0.019048	0.019489	0.019043
14	0.021435	0.019043	0.021435	0.019043

Table 4.4 Comparison for d-calls in CAC based on cutoff scheme

R_{ov}	R_{od} EV	R_{od} AV	L_{hd} EV	L_{hd} AV
1	0.007673	0.007523	0.009335	0.009229
2	0.009628	0.009503	0.011893	0.011720
3	0.011977	0.011958	0.014743	0.014847
4	0.014501	0.014415	0.018146	0.018024
5	0.016671	0.016435	0.020759	0.020681
6	0.017799	0.017787	0.022503	0.022492
7	0.018605	0.018523	0.023512	0.023494
8	0.018901	0.018853	0.023988	0.023945
9	0.018988	0.018979	0.024245	0.024111
10	0.019031	0.019023	0.024279	0.024162
11	0.019044	0.019037	0.024281	0.024176
12	0.019058	0.019042	0.024295	0.024179
13	0.019077	0.019043	0.024296	0.024179
14	0.019077	0.019043	0.024296	0.024179

Sufficiently high accuracy of the developed approximate procedures to calculating the QoS metrics under using CAC based on guard channels was observed by numerical experiments also. Since the analogous results have been shown above for

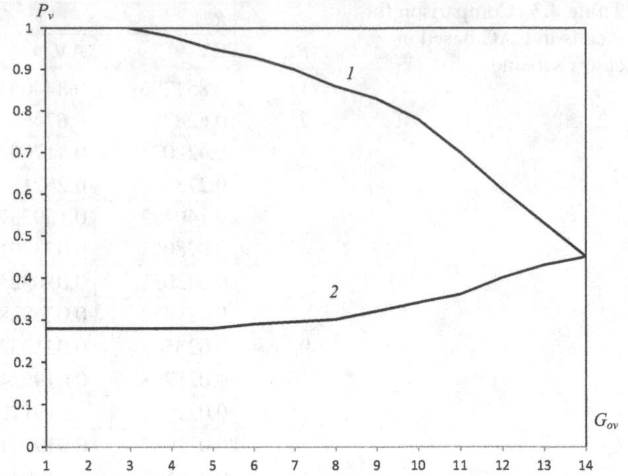

Fig. 4.7 Blocking probability of v-calls versus G_{ov}; 1—P_{ov}, 2—P_{hv}

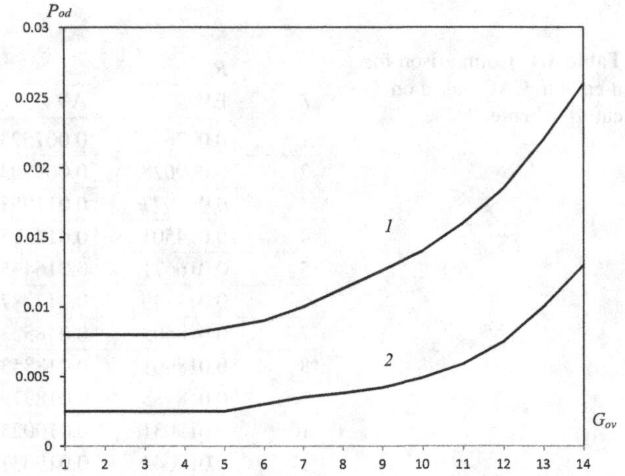

Fig. 4.8 Blocking probability of od-calls versus G_{ov}; 1—$\mu_d = 0.6$, 2—$\mu_d = 0.7$

CAC based on cutoff scheme, then due to the limited size of the book they are not presented here.

Note that the comparative analysis of QoS metrics of different CAC schemes presents a certain interest. In both strategies the total number of channels is fixed, and the variables are the parameters G_{ov} and G_{hv} (in the CAC based on guard channel scheme) and R_{ov} and R_{hv} (in the CAC based on cutoff scheme). It is evident that the behavior of QoS metrics of the model with respect to change of the above variable parameters is identical.

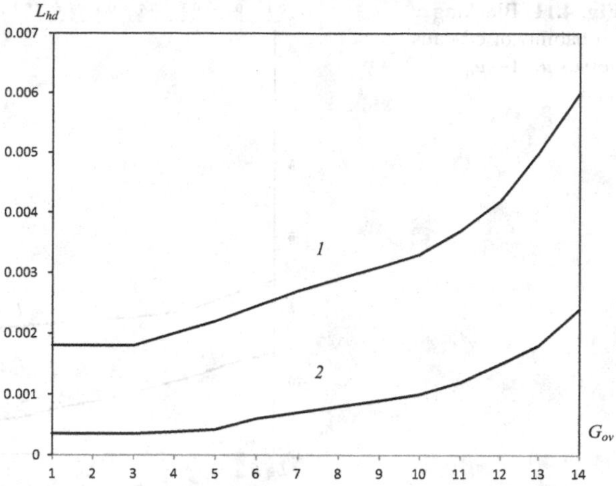

Fig. 4.9 Average length of queue of hd-calls versus G_{ov}; 1—$\mu_{\mathrm{d}}=0.6$, 2—$\mu_{\mathrm{d}}=0.7$

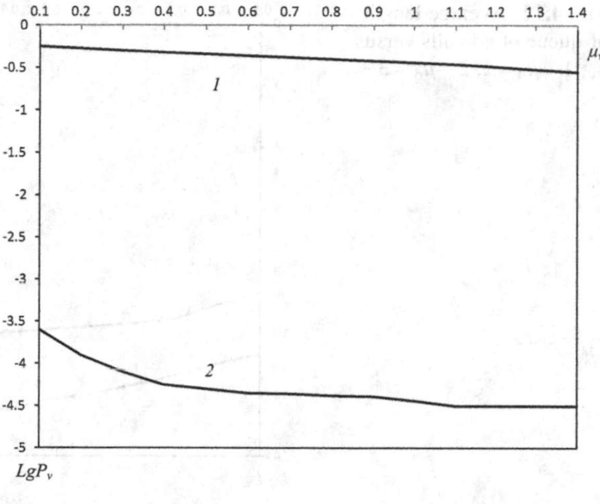

Fig. 4.10 Blocking probability of v-calls versus μ_{v}; 1—P_{ov}, 2—P_{hv}

Some comparisons are depicted in Figs. 4.13, 4.14, 4.15, and 4.16, where 1 indicates the curve of QoS metrics when using the CAC based on cutoff scheme and 2 the analogous curve when using the CAC based on guard channel scheme. The following initial data was selected: $N=20, G_{\mathrm{hv}}=14$, $G_{\mathrm{ov}}=5, \lambda_{\mathrm{ov}}=0.2, \lambda_{\mathrm{hv}}=0.1, \lambda_{\mathrm{od}}=5, \lambda_{\mathrm{hd}}=2, \mu_{\mathrm{v}}=0.1, \mu_{\mathrm{d}}=0.6$. In the described

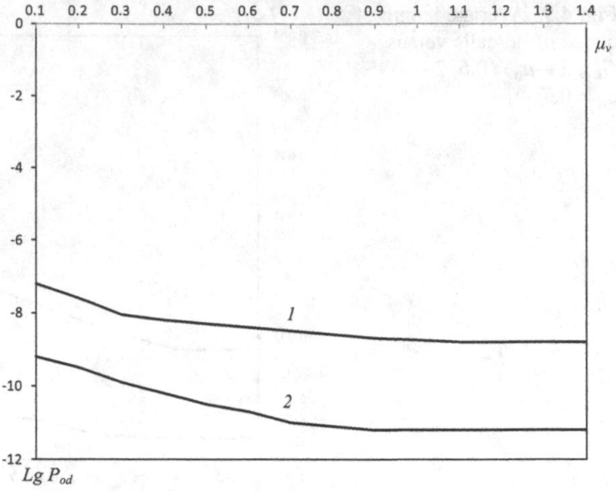

Fig. 4.11 Blocking probability of od-calls versus μ_v; 1—$\mu_d = 2$, 2—$\mu_d = 3$

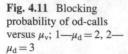

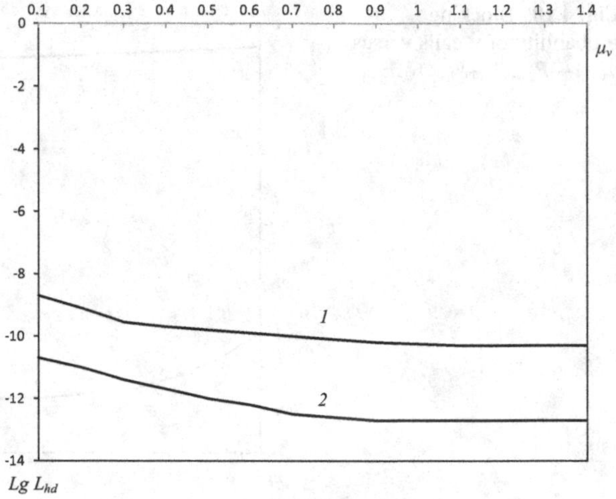

Fig. 4.12 Average length of queue of hd-calls versus μ_v; 1—$\mu_d = 2$, 2—$\mu_d = 3$

plots on the abscissa axis there is the parameter of guard channel scheme G_{ov}, and, as mentioned above, it corresponds to parameter R_{ov} of the CAC based on cutoff scheme.

From these plots one can see that for selected required initial data, the QoS metrics, which determine the level of voice call service, are much better with

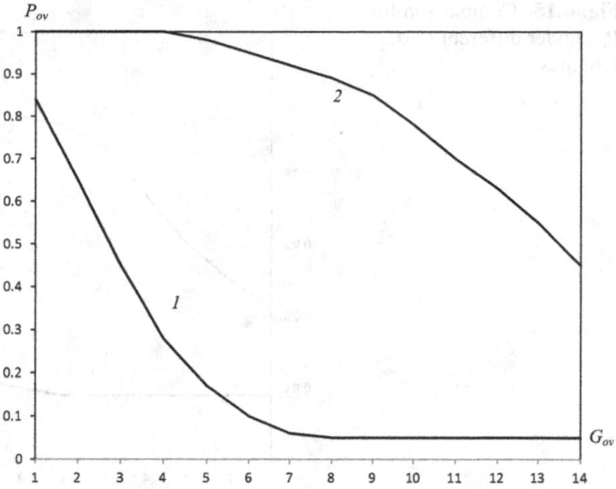

Fig. 4.13 Comparison for P_{ov} under different CAC schemes

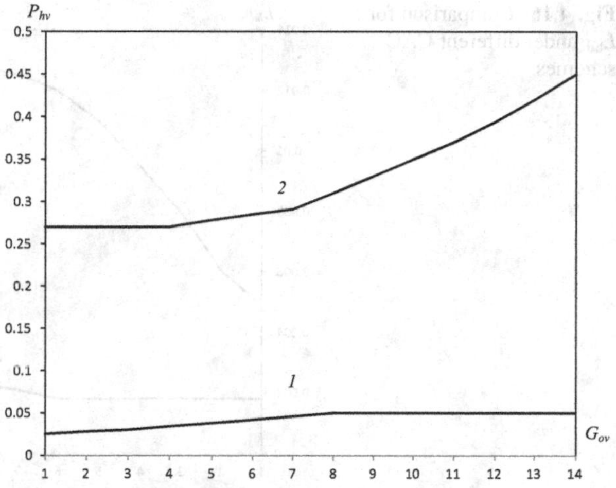

Fig. 4.14 Comparison for P_{hv} under different CAC schemes

application of cutoff scheme, and QoS metrics of data calls prefer another access scheme. However, it should be expected that with other values of initial data the guard channel scheme would be preferable for voice calls, and on the contrary, at certain initial data the cutoff scheme would be preferable for data calls.

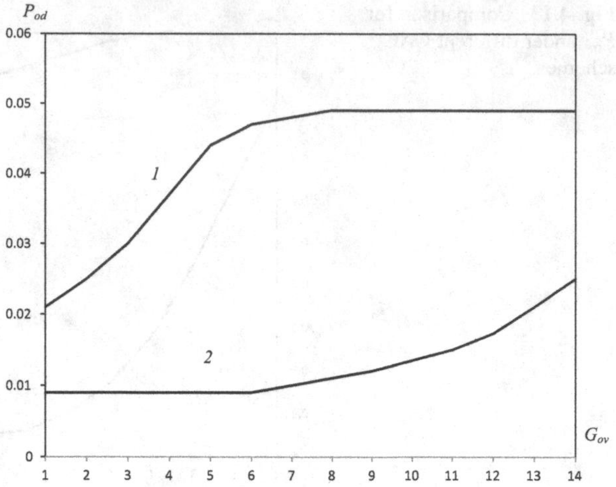

Fig. 4.15 Comparison for P_{od} under different CAC schemes

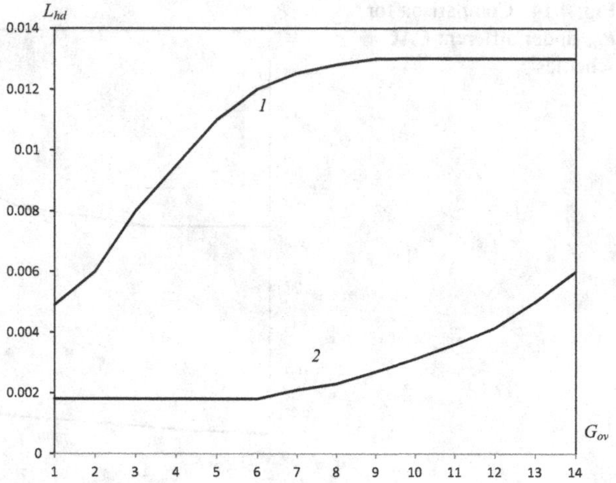

Fig. 4.16 Comparison for L_{hd} under different CAC schemes

4.2 Models with Buffering of Both New and Handover Non-Real-Time Calls

In Sect. 4.1 models with buffering only hd-calls are examined. However, in modern networks it is permitted not only buffering hd-calls but also od-calls. It is important to note that the method based on the use of balance equations is effective only for models of low and moderate dimension and cannot be used for large-scale models, especially for the models with infinite buffers. Based on these facts, there is provided an efficient numerical method for the study of such models of any dimension of the buffer, including models with an infinite buffers. Here a unified approach to the study of multiservice models (see Sect. 4.1) with finite and infinite queues of calls of both types' data is proposed. On the basis of this method the approximate method of calculating QoS metrics is developed, which has a low computational complexity for models of any dimension.

4.2.1 Integrated Access Schemes to Channels and to Buffer

Note that in this section all notations remain the same as in previous section.

First consider the model in which the access of v-calls is implemented using the cutoff scheme (see Sect. 4.1.1). However, the access of the buffered d-calls is implemented as follows:

- If upon arrival of the d-call of any type, there is at least one free channel in the system, then such call is accepted for service.
- If an incoming call belongs to the class of the od-calls and at the moment of its entry all the channels are busy, then it is taken into the buffer only when the total number of calls in the buffer is less than the value R_{od}, $1 < R_{od} \le B$, where B is a maximum buffer size (in the case of model with the finite buffer); otherwise the od-call is lost.
- If the incoming call belongs to the class of the hd-calls and at the moment of its entry all channels are busy, then it is taken to the buffer in the presence of at least one free space in the buffer; otherwise the hd-call is lost.

Upon release of any channel of the cell one d-call from the buffer is selected for transmission; any conservative queueing discipline can be used (recall that a discipline is called conservative if it does not admit idle channels in case of a queue). Since the buffer does not distinguish between types of data calls, then it is reasonable to use non-preemptive high priorities for the hd-calls.

The problem is to find QoS metrics of the system—the loss probabilities of calls of every type and average number of the d-calls in the queue, as well as their waiting time in a queue.

Let us first consider the model of a cell with the finite size of a buffer, i.e., assume that the maximum size of the buffer for queueing of the d-calls is B, where $0 < B < \infty$.

In a steady-state mode the cell state at any given time is described by the 2-D vector $\boldsymbol{n} = (n_v, n_d)$, where n_v and n_d indicate the number of voice calls in channels and total number of data calls in the system, respectively. The state space of the corresponding 2-D MC is defined as

$$S = \{\boldsymbol{n} : n_v = 0, 1, \ldots, R_{hv},\ n_d = 0, 1, \ldots, N + B;\ n_v + n_d^s \le N\}.$$

According to the introduced schemes of call access into the channels and buffer, nonnegative matrix elements of the Q-matrix of the given 2-D MC are determined from the following relations:

$$q(\boldsymbol{n},\boldsymbol{n}') = \begin{cases} \lambda_{\mathrm{v}} & \text{if } n_{\mathrm{v}} \le R_{\mathrm{ov}} - 1,\ \boldsymbol{n}' = \boldsymbol{n} + \boldsymbol{e}_1, \\ \lambda_{\mathrm{hv}} & \text{if } R_{\mathrm{ov}} \le n_{\mathrm{v}} \le R_{\mathrm{hv}} - 1,\ \boldsymbol{n}' = \boldsymbol{n} + \boldsymbol{e}_1, \\ \lambda_{\mathrm{d}} & \text{if } n_{\mathrm{d}}^{\mathrm{q}} < R_{\mathrm{od}},\ \boldsymbol{n}' = \boldsymbol{n} + \boldsymbol{e}_2, \\ \lambda_{\mathrm{hv}} & \text{if } n_{\mathrm{d}}^{\mathrm{q}} \ge R_{\mathrm{od}},\ \boldsymbol{n}' = \boldsymbol{n} + \boldsymbol{e}_2, \\ n_{\mathrm{v}}\mu_{\mathrm{v}} & \text{if } \boldsymbol{n}' = \boldsymbol{n} - \boldsymbol{e}_1, \\ n_{\mathrm{d}}^{\mathrm{s}}\mu_{\mathrm{d}} & \text{if } \boldsymbol{n}' = \boldsymbol{n} - \boldsymbol{e}_2, \\ 0 & \text{in other cases.} \end{cases} \tag{4.36}$$

In view of the proposed access scheme for voice calls one gets that QoS metrics for such kind of calls are defined similar to Eqs. (4.3) and (4.4), respectively. However, the loss probabilities of d-calls of different types are defined from the following relations:

$$P_{\mathrm{od}} = \sum_{\boldsymbol{n} \in S} p(\boldsymbol{n}) I\left(n_{\mathrm{d}}^{\mathrm{q}} \ge R_{\mathrm{od}}\right); \tag{4.37}$$

$$P_{\mathrm{hd}} = \sum_{\boldsymbol{n} \in S} p(\boldsymbol{n}) \delta\left(n_{\mathrm{d}}^{\mathrm{q}}, B\right). \tag{4.38}$$

The average number of d-calls in the queue (L_{d}) is defined similar to Eq. (4.6) where the upper bound of sum is substituted by B. So, the average waiting time in the queue of data calls is

$$W_{\mathrm{d}} = L_{\mathrm{d}} \Bigg/ \sum_{x \in \{\mathrm{od,hd}\}} \lambda_x (1 - P_x). \tag{4.39}$$

Finding the analytical solution of the appropriate SGBE is very complicated, probably unsolvable, problem. And using numerical methods of linear algebra for model with large state space is connected with the known computational difficulties. Below another approach is proposed, which allows for an approximate calculation of QoS metrics of the given system to obtain simple computational procedures, using explicit formulas.

The algorithms developed here are based on the assumptions indicated above, i.e., $\lambda_{\mathrm{d}} \gg \lambda_{\mathrm{v}}, \mu_{\mathrm{d}} \gg \mu_{\mathrm{v}}$. Under this assumption the splitting of the state space S similar to Eq. (4.7) and merging function similar to Eq. (4.8) are constructed. Note that the stationary distribution of split model with state space S_k coincides with the probability distribution of states of the model $M/M/N-k/N-k+B$ with a constant rate of service of one channel, equal to μ_{d} and state-dependent arrival intensity of the calls. The arrival intensity depends upon the state i as follows:

$$\lambda_i = \begin{cases} \lambda_{\mathrm{od}} & \text{if } i < N - k + R_{\mathrm{od}}, \\ \lambda_{\mathrm{hd}} & \text{if } i \ge N - k + R_{\mathrm{od}}. \end{cases}$$

Hence, the state probabilities of split model with state space S_k are determined as follows:

$$\rho_k(k)=\begin{cases}\dfrac{v_d^i}{i!}\rho_k(0) & \text{if } 1\le i\le N-k,\\[2ex] \dfrac{(N-k)^{N-k}}{(N-k)!}(v_d(k))^i\rho_k(0) & \text{if } N-k+1\le i\le N-k+R_{od},\\[2ex] \dfrac{v_d^{N-k}}{(N-k)!}(v_d(k))^{R_{od}}(v_{hd}(k))^{i-N+k-R_{od}}\rho_k(0) & \text{if } N-k+R_{od}< i\le N-k+B,\end{cases} \tag{4.40}$$

where $v_d(k)=v_d/(N-k)$ and $\rho_k(0)$ are found from the normalizing condition, i.e., $\sum_{i=0}^{N-k+B}\rho_k(i)=1$.

The nonnegative elements of the Q-matrix of the merged model are calculated from Eq. (4.36) by taking into account Eq. (4.40). The calculation formulas coincide with Eq. (4.12), but in this case parameters α_k (see formula (4.12)) are determined by using relations (4.40). Thus, state probabilities of merged model are defined by using formula (4.13).

For the given model approximate values of QoS metrics P_{ov}, P_{hv}, L_d and P_{hd} are determined similar to Eqs. (4.14), (4.15), (4.19), and (4.20), respectively. Note that in these formulas it is required to take into account the above-indicated fact, i.e., state probabilities of split models are determined from Eq. (4.40). The new QoS metrics—loss probability of od-calls in this model—is determined as follows:

$$P_{od}\approx\sum_{k=0}^{R_{hv}}\pi(\langle k\rangle)\sum_{i=N-k+B-R_{od}}^{N-k+B}\rho_k(i). \tag{4.41}$$

Let us consider some important special cases of the proposed integrated scheme of admission to channels and to the buffer. One of them, i.e., when there is no difference between new and handover voice calls, has been considered above (see Sect. 4.1.1). Another important special case is no difference between new and handover data calls, i.e., $R_{od}=B$. In this case the stationary distribution of the split model is determined by the following relations:

$$\rho_k(k)=\begin{cases}\dfrac{v_d^i}{i!}\rho_k(0) & \text{if } 1\le i\le N-k,\\[2ex] \dfrac{(N-k)^{N-k}}{(N-k)!}\left(\dfrac{v_d}{N-k}\right)^i\rho_k(0) & \text{if } N-k+1\le i\le N-k+B,\end{cases}$$

where $\rho_k(0)$ can be found from the normalizing condition. For this case we obtain

$$P_{hd}=P_{od}=\sum_{k=0}^{R_{hv}}\rho_k(N+B-k)\pi(\langle k\rangle).$$

It is also possible to consider a combination of these two special cases, i.e., to assume within the framework of a uniform model that $R_{ov}=R_{hv}$ and $R_{od}=B$. Then the corresponding formulas become simpler.

The proposed approach can also be used for the analysis of a model with infinite buffer, i.e., for the case $B=\infty$. In this model, as before, if all the channels are busy at the time of arrival of an od-call, the call is accepted in the buffer only if the total number of calls in the buffer is less than R_{od}, $1<R_{od}<\infty$; if the call arrived belongs to the class of hd-calls and at the time of its arrival all the channels are busy, it is always accepted in the buffer.

Applying the procedure described above, we obtain that the stationary distribution of split model with state space S_k coincides with the stationary distribution of the model $M/M/N-k/\infty$ with constant service rate of one channel, equal to μ_d, in which call arrival rate depends on the system state (see the similar model with a finite queue described above).

In split model with state space S_k, stationary mode exists if the ergodicity condition $\nu_{hd}(k)<1$ is satisfied. Since it should be satisfied for each $k=0,1,\ldots,R_{hv}$, the ergodicity condition for the original model has the form $\nu_{hd}(R_{hv})<1$. Hence, if the ergodicity condition is satisfied, the state probabilities of the split model with state space S_k are determined as follows:

$$\rho_k(k)=\begin{cases}\dfrac{\nu_d^i}{i!}\rho_k(0) & \text{if}\,1\leq i\leq N-k,\\[2ex] \dfrac{(N-k)^{N-k}}{(N-k)!}(\nu_d(k))^i\rho_k(0) & \text{if}\,N-k+1\leq i\leq N-k+R_{od},\\[2ex] \dfrac{\nu_d^{N-k}}{(N-k)!}(\nu_d(k))^{R_{od}}(\nu_{hd}(k))^{i-N+k-R_{od}}\rho_k(0) & \text{if}\,i\geq N-k+R_{od}+1,\end{cases}$$

where $\rho_k(0)$, as above, can be found from the normalizing condition.

As above, the parameters P_{ov} and P_{hv} are determined from Eqs. (4.14) and (4.15), respectively, and parameter P_{od} is calculated as follows (in this model $P_{hd}=0$):

$$P_{od}\approx 1-\sum_{k=0}^{R_{hv}}\sum_{i=0}^{N+R_{od}-k-1}\rho_k(i)\pi(\langle k\rangle).$$

After rather complex mathematical transformation, we obtain that the average queue length in this model is determined as follows:

$$L_d\approx\sum_{k=0}^{R_{hv}}\pi(\langle k\rangle)L(M/M/N-k/\infty),$$

where $L(M/M/N-k/\infty)$ denotes average queue length in the above system $M/M/N-k/\infty$ with variable call arrival rate, i.e.,

$$L(M/M/N-k/\infty)$$
$$=\rho_k(0)\frac{\nu_{\mathrm{d}}^{N-k}}{(N-k)!}\left(\sum_{i=1}^{R_{\mathrm{od}}} i(\nu_{\mathrm{d}}(k))^i+\left(\frac{\nu_{\mathrm{d}}}{\nu_{\mathrm{hd}}}\right)^{R_{\mathrm{od}}}\left(\frac{\nu_{\mathrm{hd}}(k)}{(1-(\nu_{\mathrm{hd}}(k)))^2}-\sum_{i=1}^{R_{\mathrm{od}}} i(\nu_{\mathrm{hd}}(k))^i\right)\right).$$

Now consider another restriction rules for the access of voice calls to channels based on guard channel scheme. The access of voice calls in this scheme is performed as follows:

- If at the moment of arrival of ov-call, the total number of busy channels is less than $G_{\mathrm{ov}}, 0<G_{\mathrm{ov}}<N$, then it is accepted for service; otherwise it is rejected.
- If at the moment of arrival of hv-call, the total number of busy channels is less than $G_{\mathrm{hv}}, G_{\mathrm{ov}} \le G_{\mathrm{hv}}<N$, then it is accepted for service; otherwise it is rejected.

In the given scheme as well as above d-calls are accepted according to above-described CAC scheme on the basis of reserving some places for hd-calls. Let us again first consider the model with the finite size of a buffer, i.e., assume that the maximum size of the buffer for queueing of the d-calls is B, where $0<B<\infty$.

In a stationary mode, the cell state at the arbitrary time instant is described by 2-D vector $\boldsymbol{n}=(n_{\mathrm{v}}, n_{\mathrm{d}})$, where n_{v} and n_{d} indicate the number of voice calls in channels and the total number of data calls in the cell, respectively. Hence, the state space of the given 2-D MC is

$$S=\left\{\boldsymbol{n}:\ n_{\mathrm{v}}=0,1,\ldots,G_{\mathrm{hv}},\ n_{\mathrm{d}}=0,1,\ldots,B;\ n_{\mathrm{v}}+n_{\mathrm{d}}^{\mathrm{s}}\le N\right\}.$$

The nonnegative elements of Q-matrix of the given 2-D Markov chain $q(\boldsymbol{n}, \boldsymbol{n}')$, $\boldsymbol{n}, \boldsymbol{n}' \in S$ is determined from the following relations:

$$q(\boldsymbol{n},\boldsymbol{n}')=\begin{cases}\lambda_{\mathrm{v}} & \text{if } n_{\mathrm{v}}+n_{\mathrm{d}}^{\mathrm{s}}\le G_{\mathrm{ov}}-1,\ \boldsymbol{n}'=\boldsymbol{n}+\boldsymbol{e}_1,\\ \lambda_{\mathrm{hv}} & \text{if } G_{\mathrm{ov}}\le n_{\mathrm{v}}+n_{\mathrm{d}}^{\mathrm{s}}\le G_{\mathrm{hv}}-1,\ \boldsymbol{n}'=\boldsymbol{n}+\boldsymbol{e}_1,\\ \lambda_{\mathrm{d}} & \text{if } n_{\mathrm{d}}^{\mathrm{q}}<R_{\mathrm{od}},\ \boldsymbol{n}'=\boldsymbol{n}+\boldsymbol{e}_2,\\ \lambda_{\mathrm{hv}} & \text{if } n_{\mathrm{d}}^{\mathrm{q}}\ge R_{\mathrm{od}},\ \boldsymbol{n}'=\boldsymbol{n}+\boldsymbol{e}_2,\\ n_{\mathrm{v}}\mu_{\mathrm{v}} & \text{if } \boldsymbol{n}'=\boldsymbol{n}-\boldsymbol{e}_1,\\ n_{\mathrm{d}}^{\mathrm{s}}\mu_{\mathrm{d}} & \text{if } \boldsymbol{n}'=\boldsymbol{n}-\boldsymbol{e}_2,\\ 0 & \text{in other cases.}\end{cases} \tag{4.42}$$

Loss probabilities of heterogeneous voice calls are defined similarly to Eqs. (4.22) and (4.23), and average number of d-calls in queue and average time of their waiting in queue are defined similarly to Eqs. (4.37) and (4.38), respectively.

Here, as well as above, for a finding of exact values of required QoS metrics, it is possible to use the corresponding SGBE which is made on the basis of relations (4.42). The computing difficulties specified above which arises at a finding of QoS metrics by means of an exact method are actual in the given model as well. Therefore, here it is necessary to use the approximate method developed above. Since above this method is described in detail, therefore below its some obvious stages are omitted.

In this case, a class of microstates are defined as $S_k=\{\boldsymbol{n}\in S: n_{\mathrm{v}}=k\}$, $k=0,1,\ldots,G_{\mathrm{hv}}$. From the definition of the splitting scheme of the state space, it is clear that the state probabilities of split model with state space S_k in this case are defined similarly to Eq. (4.40). The relations (4.40) and (4.42) allow one to determine the nonnegative elements of Q-matrix of the merged model:

$$q(k,k')=\begin{cases}(\lambda_{\mathrm{v}}\alpha_k+\lambda_{\mathrm{hv}}\beta_k)\rho_k(0) & \text{if } 0\le k\le G_{\mathrm{ov}}-1,\ k'=k+1,\\ \lambda_{\mathrm{hv}}(\alpha_k+\beta_k)\rho_k(0) & \text{if}\, G_{\mathrm{ov}}\le k\le G_{\mathrm{hv}}-1,\ k'=k+1,\\ k\mu_{\mathrm{v}} & \text{if}\, k'=k-1,\\ 0 & \text{in other cases,}\end{cases}$$

where $\alpha_k=\sum_{i=0}^{G_{\mathrm{ov}}-k-1} v_{\mathrm{d}}^i/i!$, $\beta_k=\sum_{i=G_{\mathrm{ov}}-k}^{G_{\mathrm{hv}}-k-1} v_{\mathrm{d}}^i/i!$.

Thus, state probabilities of the merged model are determined as

$$\pi(\langle k\rangle)=\begin{cases}\dfrac{1}{k!}\prod_{i=0}^{k-1} x_i\pi(\langle 0\rangle) & \text{if } 1\le k\le G_{\mathrm{ov}},\\ \dfrac{1}{k!}\prod_{i=0}^{G_{\mathrm{ov}}-1} x_i\prod_{j=G_{\mathrm{ov}}}^{k-G_{\mathrm{ov}}} y_j\pi(\langle 0\rangle) & \text{if}\, G_{\mathrm{ov}}+1\le k\le G_{\mathrm{hv}},\end{cases}$$

where $v_{\mathrm{v}}:=\lambda_{\mathrm{v}}/\mu_{\mathrm{v}}$, $v_{\mathrm{hv}}:=\lambda_{\mathrm{hv}}/\mu_{\mathrm{v}}$, $x_i:=(v_{\mathrm{v}}\alpha_i+v_{\mathrm{hv}}\beta_i)\rho_i(0)$, $y_i:=v_{\mathrm{hv}}(\alpha_i+\beta_i)\rho_i(0)$ and $\pi(0)$ are from the normalizing condition, i.e., $\sum_{i=0}^{G_{\mathrm{hv}}}\pi(i)=1$.

Finally, after some transformations, one obtains the following approximate formulas for the calculation of QoS metrics for d-calls (QoS metrics for v-calls are determined similarly to CAC based on cutoff scheme):

$$P_{\mathrm{od}}\approx\sum_{k=0}^{G_{\mathrm{hv}}}\pi(\langle k\rangle)\sum_{i=N-k+R_{\mathrm{od}}}^{N-k+B}\rho_k(i);$$

$$P_{\mathrm{hd}}\approx\sum_{k=0}^{G_{\mathrm{hv}}}\rho_k(N-k+B)\pi(\langle k\rangle);$$

$$L_{\mathrm{d}}\approx\sum_{k=1}^{B}k\sum_{i=0}^{G_{\mathrm{hv}}}\rho_i(N+k-i)\pi(\langle i\rangle);$$

Using the last formulas, one finds from Eq. (4.39) the average waiting time of d-calls in the queue.

Some important special cases, i.e., $G_{\mathrm{ov}}=G_{\mathrm{hv}}$ and/or $R_{\mathrm{od}}=B$, might be considered as it is investigated above. Then, the corresponding formulas are even more simplified. The proposed approach can be used to study the model with an infinite buffer, i.e., for the case $B=\infty$. Since the above similar model is described in detail, here this case is not considered.

4.2.2 Numerical Results

Let us present the results of numerical experiments for the models with limited buffers. First, consider results for CAC based on a cutoff scheme for access of voice calls.

The initial data of the model were as follows [2]: $N = 30$, $B = 50, \lambda_{ov} + \lambda_{hv} = 0.15$ call/s, $\lambda_{od} + \lambda_{hd} = 0.3$ call/s, $\mu_v^{-1} = 2$ s, and $\mu_d^{-1} = 120$ s. Thus, 30 % of the total intensity of voice calls are handover voice calls, and 80 % of the total intensity of data calls are new data calls.

For fixed values of the total number of channels and buffer size, it is possible to vary three parameters of the considered CAC scheme, i.e., R_{ov}, R_{hv} and R_{od}. In other words, the model has three degrees of freedom. Since simultaneous variation in all three parameters of the CAC scheme makes it impossible to make analytical conclusions concerning the behavior of the QoS metrics under study, in the numerical experiments described below, the parameter R_{hv} is fixed and takes the greatest possible value, i.e., $R_{hv} = 29$.

Some results of the numerical experiments for $R_{od} = 45$ are shown in Figs. 4.17, 4.18, 4.19, and 4.20. Note that an increase in one of the parameters (in the admissible domain) has a favorable effect only on QoS metrics of calls of the corresponding type. For example, in these experiments, the increase in the threshold value R_{ov} reduces the loss probability of ov-calls and increases the other QoS metrics (i.e., $P_{hv}, P_{od}, P_{hd}, L_{hd}$ and W_d). The function P_{ov} decreases with low rate (see Fig. 4.17), especially for small value of R_{ov}; however, the function P_{hv} increases rather rapidly, and the values of the functions P_{ov} and P_{hv} are equal at the point $R_{ov} = R_{hv}$. For some values of the initial data, the increase in R_{od} from 5 to 50 influences neither the behavior of these functions nor their absolute values.

The characteristic of variations in functions P_{od} and P_{hd} is identical; however, their absolute values vary within significantly different ranges (see Fig. 4.18). Similar conclusions can be made for functions L_{hd} and W_d; however, the ranges of these are almost identical in this case (see Figs. 4.19 and 4.20).

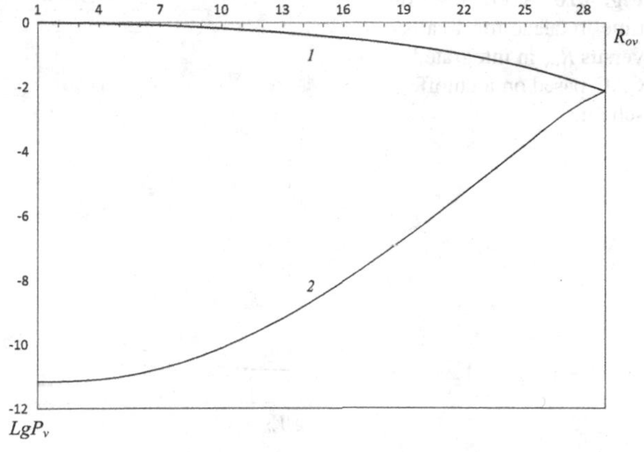

Fig. 4.17 Blocking probability of v-calls versus R_{ov} in integrated CAC based on a cutoff scheme; 1—P_{ov}, 2—P_{hv}

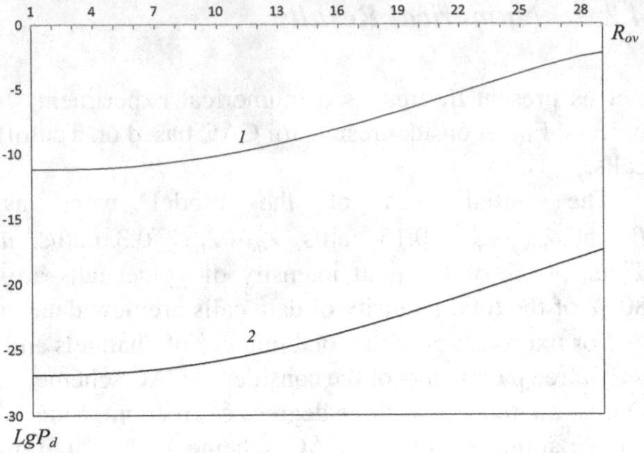

Fig. 4.18 Blocking probability of d-calls versus R_{ov} in integrated CAC based on a cutoff scheme; 1—P_{od}, 2—P_{hd}

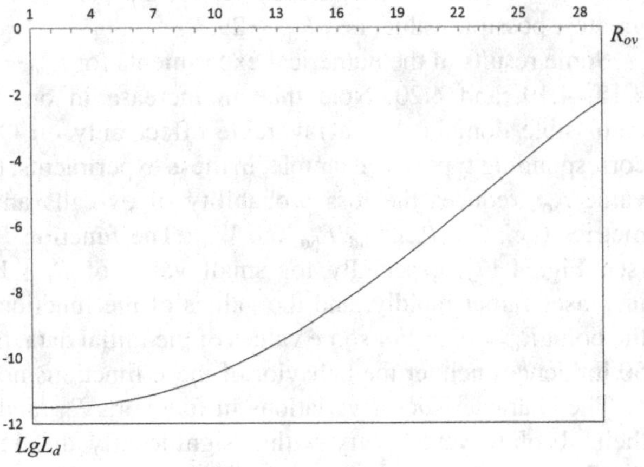

Fig. 4.19 Average length of queue of d-calls versus R_{ov} in integrated CAC based on a cutoff scheme

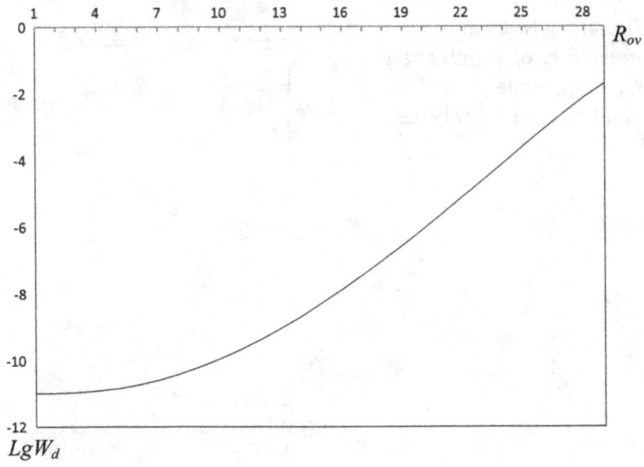

Fig. 4.20 Average waiting time in queue for d-calls versus R_{ov} in integrated CAC based on a cutoff scheme

The computational complexity of the developed approximate algorithms is essentially low than the exact ones. So, state space of the moderate-size model considered here contains 1995 states (i.e., in exact approach, SGBE contains 1995 equations), while approximate approach allows to solve the problem by means of explicit formulas (in general, the number of states is equal to $(N+B+1-(R_{hv}/2))(R_{hv}+1)$). Moreover the indicated formulas contain tabulated Erlang's formula.

The behavior of the QoS metrics versus changing the ratios of intensities of the different types of the data calls under the assumption that their sum load remains constant, i.e., $\lambda_{od}+\lambda_{hd}=0.3$, is examined. The values of other parameters of the model remain unchanged. The purpose of these numerical experiments was to study the accuracy of the proposed algorithms. The exact values of the considered QoS metrics, as in [2], are found by solving SGBE for stationary probabilities of states. Note that to solve the given SGBE, the Gauss–Seidel method has been used. Some comparisons for the initial data indicated above are shown in Tables 4.5, 4.6, and 4.7. From these tables, we conclude that along with computational simplicity, the developed approximate algorithms have also high accuracy even for models with small dimension. The performed numerical results show that accuracy of approximate formulas increases as the ratios λ_d/λ_v and μ_d/μ_v grow.

The conducted numerical experiments also allow us to determine the desirable ranges of variation of parameters of the introduced CAC schemes in order to satisfy the constraints imposed on the QoS metrics of calls of different types. For example, an absolute fair service in the sense of equal loss probabilities for voice calls of different types is observed for $R_{od}=R_{hv}$ (see Fig. 4.17). However, an absolute fair service is often not required in practice, and then it is possible to introduce the concept of ε-fair service, i.e., service where the difference between loss probabilities of voice calls of different types does not exceed a preset value $\varepsilon>0$. With the use of the numerical experiments, it is easy to solve the last problem.

Table 4.5 Comparison for v-calls in integrated CAC based on a cutoff scheme

R_{ov}	LgP_{ov} EV	LgP_{ov} AV	LgP_{hv} EV	LgP_{hv} AV
2	−0.008911	−0.008906	−1.11E+01	−1.11E+01
4	−0.035523	−0.035573	−1.09E+01	−1.10E+01
6	−0.080449	−0.080454	−1.07E+01	−1.08E+01
8	−0.140400	−0.140401	−1.03E+01	−1.04E+01
10	−0.214288	−0.214296	−9.79E+00	−9.84E+00
12	−0.303119	−0.303127	−9.28E+00	−9.19E+00
14	−0.409457	−0.409466	−8.39E+00	−8.45E+00
16	−0.537237	−0.537250	−7.61E+00	−7.62E+00
18	−0.691643	−0.691667	−6.74E+00	−6.72E+00
20	−0.878928	−0.878947	−5.79E+00	−5.77E+00
22	−1.105837	−1.105843	−4.81E+00	−4.78E+00
24	−1.379010	−1.379023	−3.79E+00	−3.78E+00
26	−1.708702	−1.708719	2.811532	2.813629
29	−2.322441	−2.322455	−1.746493	−1.744843

Table 4.6 Comparison for d-calls in integrated CAC based on cutoff scheme

R_{ov}	LgP_{ov}		LgP_{hv}	
	EV	AV	EV	AV
2	−1.11E+01	−1.12E+01	−2.69E+01	−2.69E+01
4	−1.10E+01	−1.11E+01	−2.67E+01	−2.68E+01
6	−1.09E+01	−1.08E+01	−2.64E+01	−2.65E+01
8	−1.05E+01	−1.04E+01	−2.62E+01	−2.61E+01
10	−9.88E+00	−9.89E+00	−2.59E+01	−2.56E+01
12	−9.27E+00	−9.24E+00	−2.48E+01	−2.50E+01
14	−8.48E+00	−8.50E+00	−2.45E+01	−2.42E+01
16	−7.69E+00	−7.67E+00	−2.31E+01	−2.34E+01
18	−6.71E+00	−6.77E+00	−2.23E+01	−2.25E+01
20	−5.80E+00	−5.82E+00	−2.12E+01	−2.15E+01
22	−4.81E+00	−4.83E+00	−2.07E+01	−2.05E+01
24	−3.88E+00	−3.84E+00	−1.99E+01	−1.95E+01
26	−2.912331	−2.911525	−1.88E+01	−1.85E+01
29	−1.935995	−1.936709	−1.71E+01	−1.69E+01

Table 4.7 Comparison for average length of queue of d-calls in integrated CAC based on cutoff scheme

R_{ov}	LgL_d	
	EV	AV
2	−1.14E+01	−1.15E+01
4	−1.13E+01	−1.14E+01
6	−1.10E+01	−1.11E+01
8	−1.09E+01	−1.07E+01
10	−1.01E+01	−1.02E+01
12	−9.51E+00	−9.54E+00
14	−8.79E+00	−8.80E+00
16	−7.92E+00	−7.97E+00
18	−7.03E+00	−7.07E+00
20	−6.11E+00	−6.12E+00
22	−5.10E+00	−5.13E+00
24	−4.11E+00	−4.12E+00
26	−3.09E+00	−3.12E+00
29	−1.73E+00	−1.78E+00

Now consider results for CAC based on guard channel scheme for access of voice calls. Here also the results of numerical experiments are given only for the model with a limited queue. The original model data are chosen as they indicated above.

First, consider the behavior of QoS metrics relative to changing the parameters of the adopted access scheme. Here, at the fixed values of the total number of channels and buffer size, one can change the values of three parameters of the taken

CAC (G_{ov}, G_{hv}, R_{od}); in other words, this model also has three degrees of freedom. In the following numerical experiments the parameter G_{hv} is fixed, taking the largest possible value, i.e., $G_{hv} = 29$.

As above CAC scheme, the increase in the value of one of the parameters (in the feasible region) significantly increases only the loss probability of calls of the corresponding type. Thus, in the conducted experiments, the increase of the threshold value G_{ov} reduces the probability of losing the ov-calls, while the remaining QoS metrics are increasing. However, their speeds of change are different. Indeed, the function P_{ov} decreases with sufficiently high speed (Fig. 4.21a), especially for the values G_{ov} close to the maximum possible value (i.e., G_{hv}); however, the function P_{hv} increases with a low speed at small values G_{ov} and is almost constant at its high values (Fig. 4.21b). As expected, at the point $G_{ov} = G_{hv}$ the values of the functions P_{ov} and P_{hv} are equal. It is significant to note that, as in CAC based on cutoff scheme, for the chosen values of initial data, the increase of parameter value R_{od} from 5 to 50 almost has impact neither on the behavior of these functions nor on their absolute values. These properties of the pointed functions are clearly visible especially for large values G_{ov} (here and below the curve 1—$R_{od} = 5$; curve 2—$R_{od} = 50$).

Fig. 4.21 Blocking probability of ov-calls (**a**) and hv-calls (**b**) versus G_{ov} in integrated CAC based on guard channel scheme; 1—$R_{od} = 5$, 2—$R_{od} = 50$

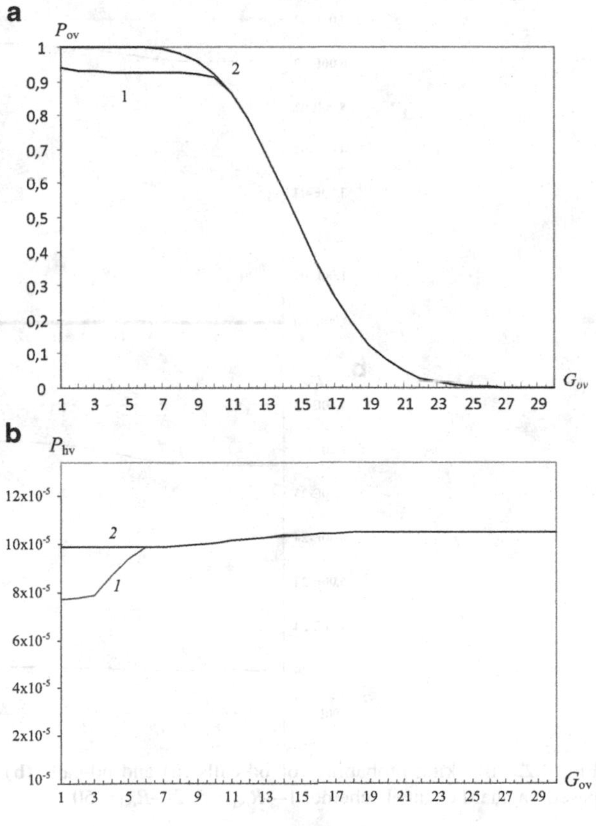

The speeds of growth of the functions P_{od} and P_{hd} are quite low, and for the values G_{ov}, close to the maximum possible value, they almost do not change (Fig. 4.22). However, their absolute values vary insignificantly in different ranges, wherein, unlike the functions P_{ov} and P_{hd}, increasing the value of the parameter R_{od} essentially affects the values of both functions. As expected, increasing the value of the parameter R_{od} has positive effects on the value of the function P_{od} (Fig. 4.22a) and the negative one on the value of the function P_{hd} (Fig. 4.22b).

The function L_d also changes with a very low speed (Fig. 4.23a), while sharply increasing the value of the parameter R_{od} almost does not affect either the nature of its changes or its absolute values. The function W_d (Fig. 4.23b) has a similar nature of the changes.

The results of numerical experiments noted above have shown that for the selected initial data, all investigated functions, except W_d, are almost constant while changing the value of the ratio $\lambda_{hd}/\lambda_{od}$. That is why the corresponding graphs are not listed here. In this case, even a tenfold increase in the value of the parameter R_{od} has little effect on the value of the function P_{od}, which is to be expected. It is easy to see that the function W_d is nonlinear, due to the nature of changes of the

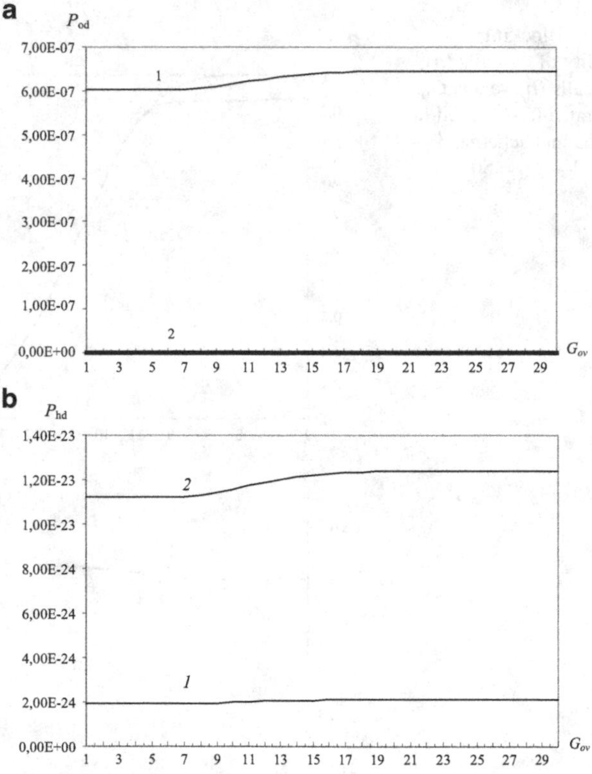

Fig. 4.22 Blocking probability of od-calls (**a**) and hd-calls (**b**) versus G_{ov} in integrated CAC based on guard channel scheme; 1—$R_{od} = 5$, 2—$R_{od} = 50$

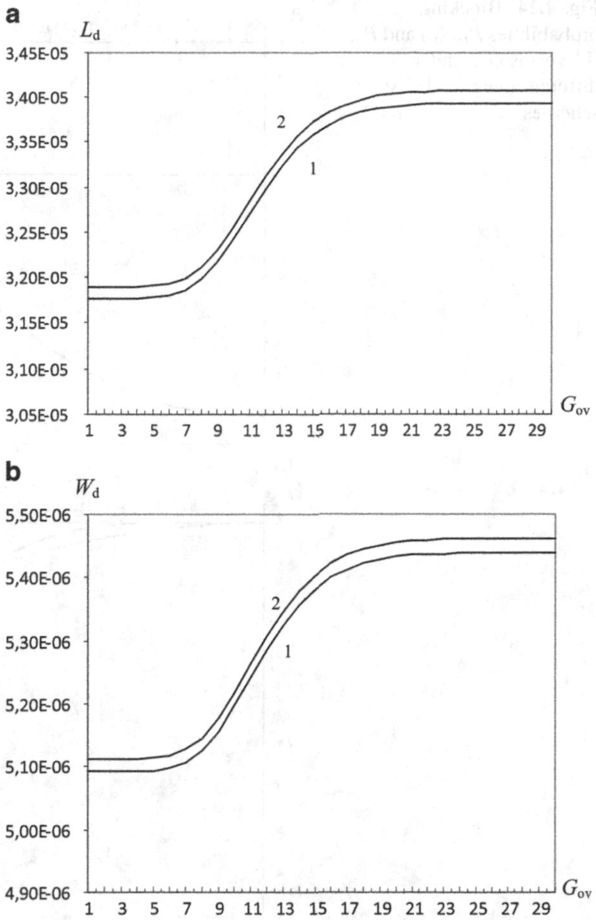

Fig. 4.23 Average length of queue of d-calls (**a**) and their average waiting time in queue (**b**) versus G_{ov} in integrated CAC based on guard channel scheme; 1—$R_{od} = 5$, 2—$R_{od} = 50$

functions P_{od} and the fact that W_d is essentially dependent on the specific values λ_{hd} and λ_{od}, but not on their relationship.

The accuracy of proposed algorithms is estimated for given CAC also. Note that for the chosen source data, the exact and approximate values of the studied QoS metrics almost coincide, and therefore, for the sake of brevity the relevant comparisons are not shown here.

We conducted a study of the behavior of the QoS metrics of the model with an infinity queue relative to changing its structural and load parameters. Such studies allow us to determine the required dimension of the buffer storage to meet the specified requirements for QoS metrics. In other words, one can answer the question whether it makes sense to use the buffer of a large enough volume for a particular system at given loads. Here let us also note that, unfortunately, the study of the accuracy of the developed formulas for models with an infinite queue cannot be done analytically. Such studies can be carried out only by simulation tools.

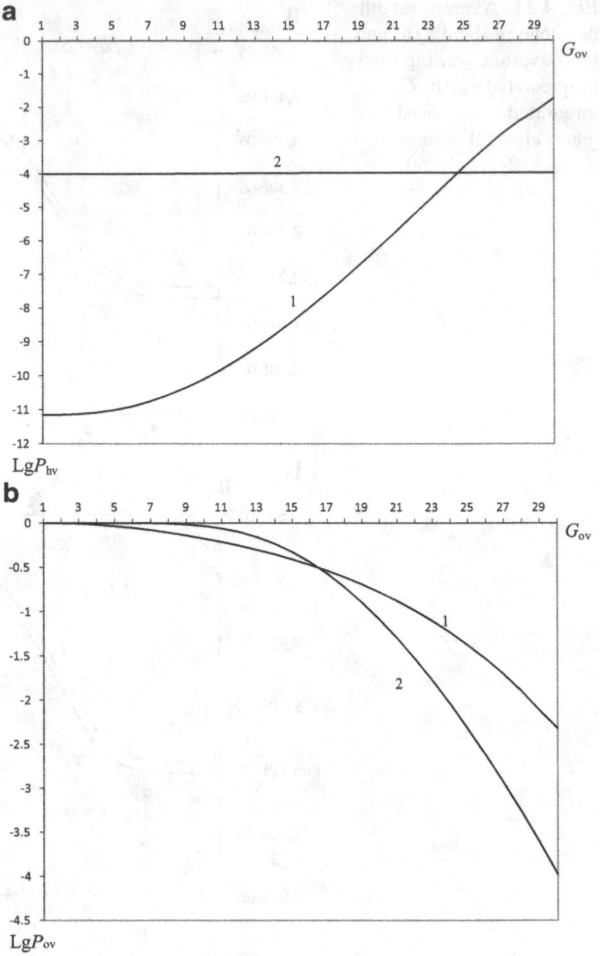

Fig. 4.24 Blocking probabilities P_{hv} (**a**) and P_{ov} (**b**) versus G_{ov} under different integrated CAC schemes

At the end of this section consider comparisons of QoS metrics in various CAC schemes. In such study we assume that load parameters are fixed. So, controllable parameters are G_{ov}, G_{hv} (in CAC based on guard channel scheme), R_{ov}, R_{hv} (in CAC based on guard channel scheme), and R_{od} (in both CAC).

In numerical experiments below the values of $G_{hv}(R_{hv})$ and R_{od} are fixed, i.e., we assume that $G_{hv} = R_{hv} = 29$ and $R_{od} = 45$. Some comparisons are shown in Figs. 4.24, 4.25, and 4.26, where labels 1 and 2 denote QoS metrics for CAC based on cutoff scheme and CAC based on guard channel scheme. The input data are the same as for the above-indicated numerical experiments. In the graphs the parameter CAC based on guard channel scheme (i.e., G_{ov}) is specified along x-axis, and as has been specified above, it corresponds to parameter R_{ov} of CAC based on cutoff scheme.

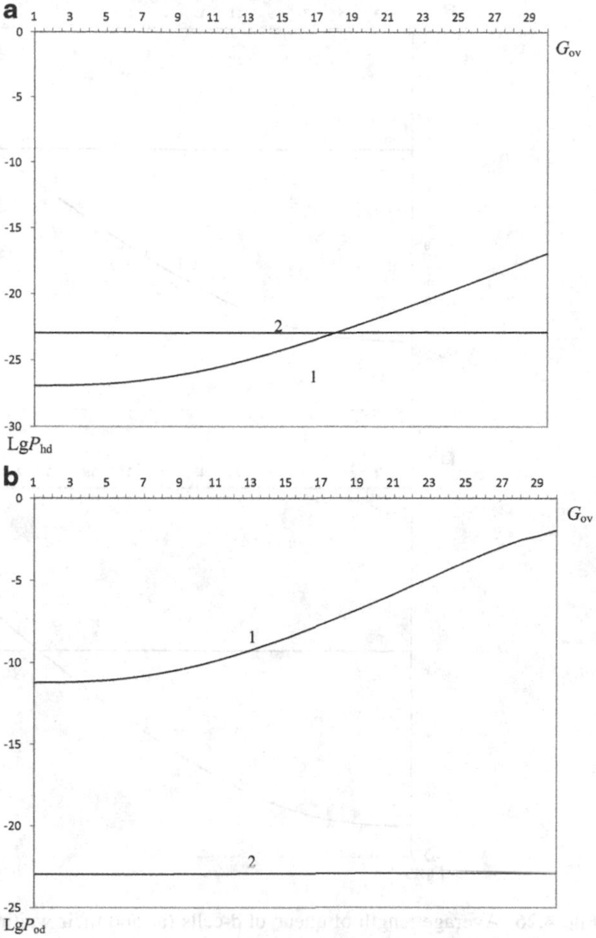

Fig. 4.25 Blocking probabilities P_{hd} (**a**) and P_{od} (**b**) versus G_{ov} under different integrated CAC schemes

From these graphs we conclude that all QoS metrics, except for the loss probability of ov-calls, in integrated CAC based on guard channel scheme for voice calls are almost linear nature, i.e., they increase with very low rates. However, all QoS metrics in integrated CAC based on cutoff scheme for voice calls have strictly nonlinear nature. In addition note that for chosen initial data, only P_{od} is essentially better under CAC based on cutoff scheme. However, quite probably, for other values of initial data for P_{od}, CAC based on guard channel scheme will be better than CAC based on cutoff scheme. In other words, the problems of searching the appropriate CAC scheme and its parameters that allow to satisfy the given QoS levels for heterogeneous calls are of certain scientific and practical interest. Various problem formulations of identifying such values of the threshold parameters of integrated CAC schemes are possible. Since in Sect. 4.1.2 similar problems and

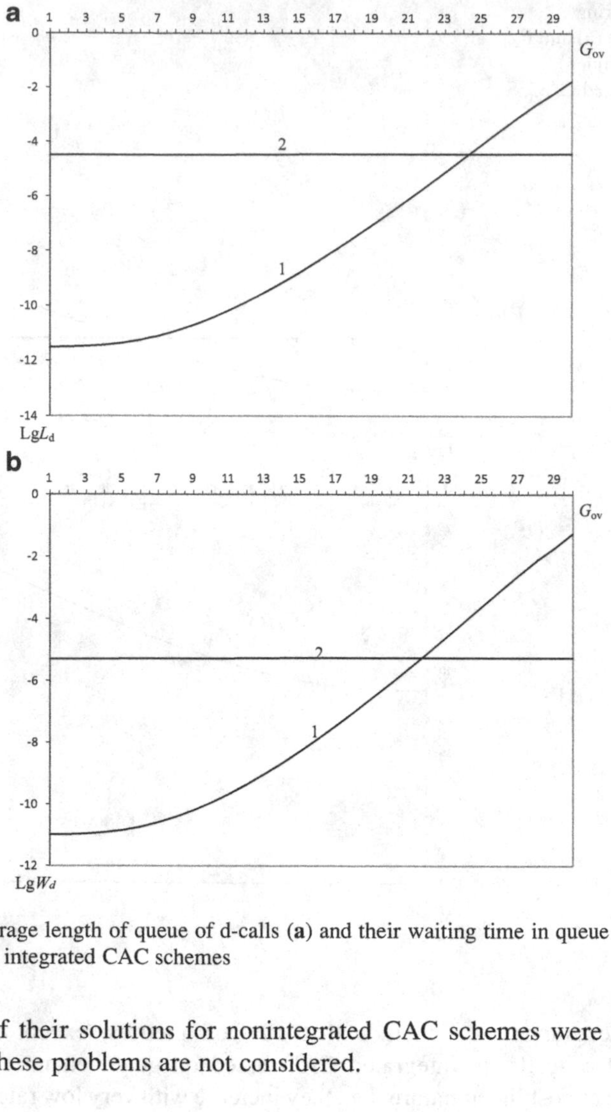

Fig. 4.26 Average length of queue of d-calls (**a**) and their waiting time in queue (**b**) versus G_{ov} under different integrated CAC schemes

algorithms of their solutions for nonintegrated CAC schemes were examined in detail, here these problems are not considered.

4.3 Conclusion

In this chapter, the models of integrated (multiservice) cellular networks in which processing of real-time calls (e.g., voice calls) and non-real-time calls (i.e., buffered data calls) takes place are studied. In such networks, data traffics are relatively tolerant to possible delays, whereas voice calls are sensitive to such delays.

A review of publications dealing with different aspects of problems related to calculation of QoS metrics in integrated cellular networks without queues has been made in works [1, 5, 15]. Note that models with queue of non-real-time calls are insufficiently investigated. Below, we give a brief consideration to the known results for models of such type.

Network models with absolute priorities of voice calls over data calls which can form a queue of limited length have been studied in [17]. Therewith, d-calls are assumed impatient, i.e., they can quit the queue without being served if their waiting time exceeds a certain random value. Iterative procedure has been developed for calculating the QoS metrics of heterogeneous calls.

The models in which interruption of data processing being under way was not allowed have been discussed in works [6, 13, 16]. The work [13] has investigated the model with full-accessible call admission control strategy for all four types of calls and infinite queue of impatient hd-calls. The model with special communication channels for v- and d-calls and common channels for their joint handling and the infinite buffer for d-calls has been studied in the work [16]. The analogous model with the finite buffer has been studied in [6], the approximate results being obtained from the corresponding results of the work [16] by applying the known cutoff procedure of distribution tail of a queue length. In the works [6, 13, 16] for calculating metrics of QoS, one employs the 2-D generating function method. However, as mentioned by authors, such approach is cumbersome and unconstructive since it does not allow one to develop the efficient calculation procedures for solving a stated problem even for models of small dimension.

In models with queues of non-real-time calls along with strategy of access to channels, it is necessary to define also the strategy of access to the buffer. So, in [2] buffering both new and handover data calls is admitted, while in [8] and [9] buffer is assigned only for handover data calls. Thus, in [2] and [9], CAC strategy to channels is based on the guard channels, and in [8] CAC strategy uses the cutoff scheme. Since in [8] and [9] buffering is allowed only hd-calls, CAC strategy to the buffer is defined by simple scheme, i.e., any arrived hd-call is accepted if the buffer is not full. In [2], CAC strategy to buffer is controlled by the scheme based on the reservation of some capacity of the buffer for handover data calls. Note that the method offered in [2] to calculate QoS metrics is based on the use of the system of balance equations. It is effective only for models of small and moderate dimension of the buffer and cannot be applied to models with the large dimension of the buffer, especially for models with infinite queues. An approximate method for solving the above-indicated problems for the model with queues (finite and infinite) of both new and handover data calls is proposed in [10–12]. The approximate calculation method has comparatively low computation complexity for models of arbitrary dimension. The proposed approximate method can also be used to analyze the models with impatient and wideband data calls.

Models of integrated voice/data wireless network with finite common queue of real-time calls (i.e., new and handover voice calls) are considered in [4]. In this work, fixed guard channel scheme is used (i.e., exclusive channels for handover voice calls are fixed), and there is no difference between new and handover data calls. It is assumed that channel holding times of the non-guard and guard channels have different means and a recursive approach to calculate the steady-state probabilities of appropriate 2-D MC is developed.

References

1. Ahmed MH (2005) Call admission control in wireless network. A comprehensive survey. IEEE Commun Surv Tutorials 7(1):50–69
2. Carvalho GHS, Martins VS, Frances CRL, Costa JCWA, Carvalho SV (2008) Performance analysis of multi-service wireless network: An approach integrating CAC, scheduling, and buffer management. Comput Electr Eng 34:346–356
3. Das SK, Jayaram R, Kakani NK, Sen SK (2000) A call admission and control scheme for QoS provisioning in next generation wireless networks. Wirel Netw 6:17–30
4. Feng W, Kowada M (2008) Performance analysis of wireless mobile networks with queueing priority and guard channels. Int Trans Oper Res 15:481–508
5. Ghaderi M, Boutaba R (2006) Call admission control for voice/data integration in broadband wireless networks. IEEE Trans Mobile Comput 5(3):193–207
6. Haung YR, Lin YB, Ho JM (2000) Performance analysis for voice/data integration on a finite-buffer mobile system. IEEE Trans Veh Tech 49(2):367–378
7. Kolmogorov A (1936) Zum theorie der Markoffschen ketten. Mathematische Annalen B112:155–160
8. Melikov AZ, Ponomarenko LA, Kim CS (2011) Methods for analysis of multi-service wireless cellular communication networks with queues of non-real time calls. Part I. Cut-off strategy. J Autom Inform Sci 43(7):46–60
9. Melikov AZ, Ponomarenko LA, Kim CS (2011) Methods for analysis of multi-service wireless cellular communication networks with queues of non-real time calls. Part II. Guard channels strategy. J Autom Inform Sci 43(10):19–28
10. Melikov AZ, Ponomarenko LA, Kim CS (2013) Analysis of a cellular network model with multi-parameter control for call admission to channels and to data call buffer. Cybern Syst Anal 49(6):836–844
11. Melikov AZ, Ponomarenko LA, Kim CS (2013) Approximate analysis method for multi-service cellular networks with the buffer for data calls. J Autom Inform Sci 45(6):13–24
12. Melikov AZ, Ponomarenko LA, Kim CS (2014) Unified approach to analysis of models of cellular networks with buffering of data calls. Autom Rem Contr 75(8):1420–1432
13. Pavlidou FN (1994) Two-dimensional traffic models for cellular mobile systems. IEEE Trans Commun 42(2/3/4):1505–1511
14. Ponomarenko L, Kim CS, Melikov A (2010) Performance analysis and optimization of multi-traffic on communication networks. Springer, Heidelberg
15. Wei L, Chao X (2004) Modeling and performance evaluation of cellular mobile networks. IEEE/ACM Trans Netw 12(1):131–145
16. Yuang MC, Haung YR (1998) Bandwidth assignment paradigm for broadband integrated voice/data networks. Comput Commun 21(3):243–253
17. Zhuang W, Bensaou B, Chua KC (2000) Handoff priority scheme with preemptive finite queuing and reneging in mobile multiservice networks. Telecommun Syst 15(1–2):37–51

Chapter 5
Priority Schemes in Packet Switching Networks

In this chapter, new queueing models of switching equipment in packet networks and corresponding algorithms necessary for their analyzing are developed. Traditional approaches to solving the indicated problems focused mainly on simple queueing models with single traffic and do not take into account various QoS requirements of heterogeneous cells (packets). However, nowadays we have to consider the models of switch that supported both delay-sensitive applications (such as real-time voice calls) and delay-insensitive applications (such as non-real-time data calls). Note that usually delay-insensitive applications are loss sensitive, while delay-sensitive applications have some tolerance to loss. In other words, in unique model, it is necessary to take into consideration various QoS requirements for heterogeneous cells which contradict each other.

Here both kinds of models of shared memory switches with typed and common output ports are considered. For these models state-dependent priority schemes are introduced. In models of the switch with typed output ports only space priorities are defined, while for models of switch with common output ports the multiple space and time priorities are used. Moreover, models of switches with jump priorities are investigated in detail. Note that in all models there are some controllable parameters which allow to regulate the values of QoS metrics. Both exact and approximate approaches to calculate and optimize the QoS metrics of considered switches are developed.

5.1 Space and Time Priorities in Switch Modeling

First of all, note that within this chapter we will use the terms "user," "application," "connection," or "flow" to describe the traffic carried by the switch. And also the terms "packet" and "cell" are used as synonyms.

The main building block of high-performance packet networks is the switching equipment which connects the several incoming and outgoing links together.

A. Melikov, L. Ponomarenko, *Multidimensional Queueing Models in Telecommunication Networks*, DOI 10.1007/978-3-319-08669-9_5

In other words, the main function of the switch is to transfer the data flows from one incoming link to outgoing link. To determine the proper outgoing link for each incoming packet switch, use the information in its header. A switch is a hardware device that contains several components: a network processor unit, a controller, an input/output ports, and switch fabric. Here we do not consider in more details the review to known switch design, but we study the basic performance measures of the switch. Among them more important are two QoS metrics: packet loss and packet delay. Both QoS metrics essentially influence on performance of the entire packet switching network (PSN).

The switch-limited resources being shared among the users are available output link bandwidth and local buffer space. In order to realize the appropriate resource sharing schemes, a scheduling algorithm must be implemented at each network switch. The goal of scheduler is to provide the given QoS level for different types of flows going through the switch. To reach this goal the scheduler in the switch performs two main functions: (1) select packet from the buffer (queue) for forwarding to a certain switch output; (2) select packet from the buffer for dropping during the periods of congestion.

As it was noted above, in modern PSN different service classes are supported by the switch, and they do not require an equal share of bandwidth and buffer. It means that the scheduler must allocate the limited switch resources in some optimal (suboptimal or fair) manner so that it satisfies the desired QoS level for each class of service. In other words, the scheduling problem is an optimization problem.

Before study the scheduler functions consider traffic aggregation problems in PSN. In order to reach the desired QoS level, the scheduler has to keep state information about each active flow, i.e., a scheduling algorithm must operate on a per-flow basis. However, the large numbers of states that must be processed lead the scheduler to slow down and thus limit the number of packets that can be accepted. In other words, the scheduler must be easy to implement in switch especially at high-speed networks, i.e., computations to be done by the scheduler to be small in number and simple to calculate. It means that effective schedulers should aggregate several traffics into broad service classes to reduce the amount of state information and workload. In order to carry out aggregation heterogeneous traffics, it is necessary take into account their common properties. By considering the QoS ratios between the service classes and their physical properties, it is expedient that packets in PSN divide into two classes—real-time packets and non-real-time packets. This approach is promising since it deals with large aggregates of network traffic and thus mathematical analysis of switches becomes simpler. Note that in this aggregation procedure the scheduler loses specific information about the status of each user and thus guarantees can be provided on the QoS metrics for aggregated groups of users, but each user cannot be guaranteed specific QoS level. Below, we will investigate the proposed aggregation procedure that does not require full per-flow state information since it is more likely to be implemented in practice.

Now consider each scheduler functions separately. Packet selecting scheme for forwarding to a switch output becomes actual when several packets in buffer will request to access a certain output port. In such situations, the scheduler must decide which stored packet must be sent next. Different selection schemes could be implemented depending on the switch architecture (i.e., location of the buffers within switch and output port assignment) and service classes of the queues. For example, for switch with typed output ports (i.e., switch in which output ports are divided between service classes), this problem might be solved trivially. In such switch there are no conflict situations within same flow of packets. Absolutely other situation arises for switch with common output ports since in such switch it is necessary to solve the nontrivial problems.

To solve problems that arise in packet selecting from buffer of switch with common output ports, various time priorities (TP) are used. As it was mentioned above, real-time applications have rigid constraints on delay, while non-real-time applications are handled using best-effort packet transfer policy with no delay guarantees. So, real-time applications have high priority compared to non-real-time applications to select the packet for forwarding to a switch output. However, such static TP assignment is not a very effective one since in this case non-real-time applications will be waiting in buffer for a long time. It means that state-dependent (or time-dependent) TP should be used to satisfy the given QoS level of heterogeneous users.

Another scheduler function is determining the packet dropping scheme. This function is implemented by space priorities (SP) in switches for which common buffer is shared among different users. Space priorities might be divided into two broad classes: reactive SP and preventive SP. The class of reactive SP contains priority schemes which decide to drop the packet during period of congestion, i.e., when buffer has become full and new packet has arrived. In such case the scheduler should decide which packets to drop (in literature this kind of SP is called tail-dropping scheme also). Usually reactive SP uses the push-out schemes, i.e., one type of packet might push out the other type of packet from the buffer and takes its replace. Unlike the reactive SP, preventive space priorities belong to the class of early drop schedulers where a packet might be dropped even when the buffer is not full. In other words, in preventive SP, the switch drops arriving packet when the queue size (separate or total) exceeds a certain state-dependent threshold. It is clear that preventive SP is based on non-push-out scheme. Note that space priorities might contain some probabilistic parameters.

It is important to note that application of high-priority scheme for one type of traffic (either for real-time type or non-real-time type) in switches of multimedia networks is unacceptable since lower priority traffic will have both large delay and high loss which does not satisfy the QoS requirements of the mentioned traffics. This is why we need multiple priorities in switches such as high TP and low SP for delay-sensitive (i.e., real-time) traffic and high SP and low TP for loss-sensitive (non-real-time) traffic. Here we again underline that any priority scheme in the scheduling algorithm must be easy to implement and does not require a huge computations.

The placement of the buffers in the switch has direct impact on the overall switch performance. There are many switches buffering architecture. Three main switch types are the following: input queueing switch, output queueing switch, and shared buffer switch. Advantages and disadvantages of each buffer design schemes are well known and here we do not discuss these issues. Below we consider models of shared buffer (memory) switches with a single common buffer to store all arriving packets.

5.2 Models of Switches with Typed Output Ports

In high-speed packet switching networks, different users impose different requirements on the QoS metrics. As it was mentioned above, real-time traffics place rigid requirements regarding potential delays, while for non-real-time data the number of lost packets is an extremely important factor. To meet these requirements, effective utilization of the network resources, i.e., the buffer space and transmission bands, is essential. Obviously, the QoS metrics of users depend largely on the load of the flows, the dimension of the buffer, as well as the particular distribution scheme adopted for the common network resources. It is assumed that the load of the data flows is irregular and that, therefore, any improvement in QoS metrics is possible only through management of the buffer, i.e., by varying the buffer dimension and the buffer sharing scheme. The selection of an efficient buffer sharing scheme is important, since a trivial technique for improving QoS metrics through increasing the buffer dimension is not always an acceptable procedure for solving the problem.

Buffer allocation schemes may be divided into two classes: push-out (PO) schemes and non-push-out (NPO) schemes. When push-out schemes are employed, packets of any type are received into the buffer when there are free spaces present, and, in certain cases, a newly arrived packet may take the place of another type of packet in a completely full buffer. Such substitution is called push-out. If a non-push-out scheme is employed, there is the guarantee that a packet which has been accepted into the buffer will be transmitted through the appropriate outgoing port. Note that non-push-out schemes are simpler to implement in practice.

In this section we are considering SP based on both classes of buffer allocation schemes for the switch in which the outgoing ports are specialized relative to type of packet. The basic problems in the study of buffer sharing schemes are known as the "tyranny of dimension." Such a situation occurs when there are a large number of different types of packets and/or a high-dimension buffer. In this case, the problem is complicated even further by the fact that in the presented schemes, unlike the other well-known sharing schemes, a multiplicative solution for appropriate multidimensional MC does not exist. This means that under such assumptions, the only way of investigating such a scheme is to solve SGBE of very large dimension (an alternative approach is to perform simulation modeling). Below to overcome these difficulties an approximate approach is proposed.

5.2.1 *Space Priorities Based on Non-Push-Out Schemes*

First, consider the space priorities based on NPO scheme; more particularly, we study partial sharing (PS) scheme of the common buffer. In this scheme, some threshold for packets of a definite type is established. In other words, once the total number of different types of packets in the buffer has reached the threshold value, new packets of the same type are lost (blocked), while packets of another type are accepted until the buffer is completely full. The main advantage of this scheme is that, first, it is a generalization of the well-known and commonly accepted complete sharing (CS) scheme; hence its characteristics will be no worse than those of the CS scheme; second, it possesses an element of adaptability due to the threshold parameter, and hence, optimization problems for this type of scheme with respect to some selected QoS metrics may be formulated and solved.

The model description consists in the following. A buffer space of dimension B is used simultaneously by two types of packets, and the outgoing ports are specialized relative to type of packet, i.e., only packets of type i are transmitted through port i, $i = 1, 2$. The process of arrival of packets of type i obeys a Poisson law with parameter λ_i, while the packet service time is an exponentially distributed random variable with mean μ_i^{-1}, $i = 1, 2$. A packet of either type frees its place in the buffer only after its transmission is complete, i.e., during the period it is serviced in the buffer, it continues to occupy space in the buffer.

The PS scheme is realized in the following way. Packets of type 1 are received only when at the time they arrive, the total number of packets of both types is less than some specified number r, $0 < r \leq B$; otherwise these packets are lost. Packets of type 2 are lost only after the buffer is completely full, i.e., they are always received into the buffer if there is at least a single free space in the buffer.

Note 5.1 Complete sharing scheme is a special case of a PS scheme, i.e., when $r = B$, the present scheme coincides with the CS scheme.

The problem is to develop an effective approach to the calculation and optimization of the QoS metrics of the PS scheme. The main QoS metrics are cell loss probability (CLP) and cell transfer delay (CTD).

By virtue of the above assumptions concerning the type of distribution functions governing outgoing flows and their processing times, the performance of the switch is described by a 2-D MC with states of type $\boldsymbol{n} = (n_1, n_2)$, where n_i indicates the number of packets of type i in the system, $i = 1, 2$. The state space of this switch is specified in the following form:

$$S = \{\boldsymbol{n} : n_1 = 0, 1, \ldots, r; n_2 = 0, 1, \ldots, B; n_1 + n_2 \leq B\}. \tag{5.1}$$

The nonnegative elements of the Q-matrix of the 2-D MC are determined as follows:

$$q(\boldsymbol{n},\boldsymbol{n}')=\begin{cases}\lambda_1 & \text{if } n_1+n_2<r,\ \boldsymbol{n}'=\boldsymbol{n}+\boldsymbol{e}_1,\\ \lambda_2 & \text{if } \boldsymbol{n}'=\boldsymbol{n}+\boldsymbol{e}_2,\\ \mu_i & \text{if } \boldsymbol{n}'=\boldsymbol{n}-\boldsymbol{e}_i,\ i=1,\ 2,\\ 0 & \text{in other cases.}\end{cases} \tag{5.2}$$

The stationary probability of state $\boldsymbol{n}\in S$ is denoted $p(\boldsymbol{n})$ (the existence of a stationary mode follows from the finiteness and irreducibility of the particular 2-D MC).

The QoS metrics of the model are defined in terms of its stationary distribution. So, the basic QoS metrics—the probability that cells of type i are lost $\text{CLP}_i(B,r)$, $i=1,2$, are determined from the following formulas:

$$\text{CLP}_1(B,r)=\sum_{\boldsymbol{n}\in S}p(\boldsymbol{n})I(n_1+n_2\geq r); \tag{5.3}$$

$$\text{CLP}_2(B,r)=\sum_{\boldsymbol{n}\in S}p(\boldsymbol{n})\delta(n_1+n_2,B). \tag{5.4}$$

The average cell transfer delay for the cells of type i is denoted by $\text{CTD}_i(B,r)$, $i=1,2$. These parameters are calculated by modified Little's formula:

$$\text{CTD}_i(B,r)=\frac{Q_i(B,r)}{\lambda_i(1-\text{CLP}_i(B,r))}, \tag{5.5}$$

where $Q_i(B,r)$ is the average number of cells of type i in the buffer, $i=1,\ 2$. Last parameters are calculated as follows:

$$Q_i(B,r)=\sum_{k=1}^{x_i}k\xi_i(k), \tag{5.6}$$

where $x_i=\begin{cases}r & \text{if } i=1,\\ B & \text{if } i=2;\end{cases}$ $\xi_i(k)=\sum_{\boldsymbol{n}\in S}p(\boldsymbol{n})\delta(n_i,k)$, $i=1,\ 2$ are marginal distributions of the initial model.

State probabilities are determined from appropriate SGBE which is constructed by using relations (5.2). Note that when PS scheme is used to buffer allocation, there will not exist a multiplicative solution for the stationary distribution of the given 2-D MC. Indeed, from relations (5.2), we conclude that the investigated 2-D MC is not reversible, i.e., there exists the transition $(n_1,n_2)\rightarrow(n_1-1,n_2)$ with intensity μ_1 where $n_1+n_2\geq r$, but the inverse transition does not exist. That is, to determine the QoS metrics (5.3)–(5.5), it is necessary in this case to solve an SGBE each time for specific values of the structure and load parameters of the system. This step entails enormous computational difficulties if the state space (5.1) has large dimension.

Below, we propose an approximate approach as a way of overcoming these difficulties. It is assumed that cells of some type arrive at a higher rate and that they are also serviced more rapidly than cells of another type. To make the discussion concrete, below we assume that $\lambda_2 >> \lambda_1$ and $\mu_2 >> \mu_1$. Note that such a performance mode is the ordinary mode for multimedia networks in which the load parameters of voice cells somewhat exceed the corresponding parameters of cells of video information.

Let us consider the following splitting of the state space S:

$$S = \cup_{i=0}^{r} S_i,\ S_i \cap S_j = \varnothing, \tag{5.7}$$

where $S_i = \{\boldsymbol{n} \in S : n_1 = i\}$, $i = 0, 1, \ldots, r$. In other words, in splitting Eq. (5.7), subset S_i contains those states of S in which the number of cells of type 1 is equal to i.

Note 5.2 The accepted assumption ensures realization of the requirement necessary for correct application of algorithms of state space merging. Indeed, the assumption ensures small transition probabilities between states which belong to different classes in splitting Eq. (5.7), as compared with respect to transition probabilities between the states inside the classes.

Further, the classes of states S_i, $i = 0, 1, \ldots, r$ are combined into individual merged states $\langle i \rangle$, and a merge function on the state space S is introduced:

$$U(\boldsymbol{n}) = \langle i \rangle \text{ if } \boldsymbol{n} \in S_i, \quad i = 0, 1, \ldots, r. \tag{5.8}$$

Thus, the merge function (5.8) defines a merged model, which is one-dimensional birth-and-death process (1-D BDP) with state space $\Omega = \{\langle i, \rangle, :, i, = 0, 1, , , \ldots, , , r\}$. To construct the Q-matrix of a merged model, it is necessary to determine the stationary distribution within each split model with state space S_i, $i = 0, 1, \ldots, r$.

The probability of state $(i,j) \in S_i$ is denoted by $\rho_i(j)$. In light of Eq. (5.2), it is defined as the stationary distribution of a 1-D BDP. We will distinguish two cases:

Case $v_2 \neq 1$, where $v_2 = \lambda_2/\mu_2$:

$$\rho_i(j) = v_2^j \frac{1 - v_2}{1 - v_2^{B+1-j}}. \tag{5.9}$$

Case $v_2 = 1$:

$$\rho_i(j) = \frac{1}{B + 1 - i}, \quad i = 0, 1, \ldots, r : j = 0, 1, \ldots, B - i. \tag{5.10}$$

Then, from Eqs. (5.2), (5.9), and (5.10), we conclude that the elements of the

Q-matrix of the merged model, which is 1-D BDP, are determined in the following way:

Case $v_2 \neq 1$:

$$q(\langle i\rangle, \langle j\rangle) = \begin{cases} \lambda_1 \dfrac{1 - v_2^{r-i}}{1 - v_2^{B+1-i}} & \text{if } j = i + 1, \\ \mu_1 & \text{if } j = i - 1, \\ 0 & \text{in other cases.} \end{cases} \tag{5.11}$$

Case $v_2 = 1$:

$$q(\langle i\rangle, \langle j\rangle) = \begin{cases} \lambda_1 \dfrac{r - i}{B - i + 1} & \text{if } j = i + 1, \\ \mu_1 & \text{if } j = i - 1, \\ 0 & \text{in other cases.} \end{cases} \tag{5.12}$$

Using conditions (5.11) and (5.12), we find the stationary distribution $\pi(\langle i\rangle)$, $\langle i\rangle \in \Omega$ of the merged model:

Case $v_2 \neq 1$:

$$\pi(\langle i\rangle) = v_1^i \prod_{j=0}^{i-1} C(j)\pi(\langle 0\rangle), \quad i = 0, 1, \ldots, r, \tag{5.13}$$

where $v_1 = \lambda_1/\mu_1$, $C(j) = (1 - v_2^{r-j})/(1 - v_2^{B-j+1})$.

Case $v_2 = 1$:

$$\pi(\langle i\rangle) = v_1^i \prod_{j=0}^{i-1} D(j)\pi(\langle 0\rangle), \quad i = 0, 1, \ldots, r, \tag{5.14}$$

where $D(j) = (r - j)/(B - j + 1)$.

In both cases, $\pi(\langle 0\rangle)$ is determined from normalizing condition, i.e., $\sum_{i=0}^{r} \pi(\langle i\rangle) = 1$.

Now the stationary distribution $p(i,j)$, $(i,j) \in S$ of the initial model may be approximately determined as follows:

$$p(i,j) \approx \rho_i(j)\pi(\langle i\rangle). \tag{5.15}$$

Finally, using Eqs. (5.9)–(5.15) we find the following approximate formulas to calculate the QoS metrics:

Case $v_2 \neq 1$:

$$\mathrm{CLP}_1(B,r) \approx \left(1 - v_2^{B-r+1}\right)\sum_{i=0}^{r} \frac{v_2^{r-i}}{1 - v_2^{B-i+1}}\pi(\langle i\rangle). \tag{5.16}$$

$$\mathrm{CLP}_2(B,r) \approx \sum_{i=B-r}^{B} L(v_2, i)\pi(\langle B-i\rangle). \tag{5.17}$$

$$Q_1(B,r) \approx \sum_{i=1}^{r} i v_1^i \prod\nolimits_{j=0}^{i-1} C(j)\pi(\langle 0\rangle). \tag{5.18}$$

$$Q_2(B,r) \approx (1-v_2)\left(\sum_{i=0}^{r} \frac{\pi(\langle i\rangle)}{1 - v_2^{B-i+1}}\sum_{j=1}^{B-r} j v_2^j + \sum_{i=B-r+1}^{B} i v_2^i \sum_{j=0}^{B-i} \frac{\pi(\langle j\rangle)}{1 - v_2^{B-j+1}}\right). \tag{5.19}$$

Hereinafter, $L(v,k)$ denotes the stationary probability of a loss in the classical queueing system $M/M/1/k+1$ with load v Erl, i.e., $L(v,k) = v^k(1-v)/(1-v^{k+1})$.

Case $v_2 = 1$:

$$\mathrm{CLP}_1(B,r) \approx (B-r+1)\sum_{i=0}^{r} \frac{\pi(\langle i\rangle)}{B-i+1}. \tag{5.20}$$

$$\mathrm{CLP}_2(B,r) \approx \sum_{i=B-r}^{B} \frac{\pi(\langle B-i\rangle)}{i+1}. \tag{5.21}$$

$$Q_1(B,r) \approx \sum_{i=1}^{r} i v_1^i \prod\nolimits_{j=0}^{i-1} D(j)\pi(\langle 0\rangle). \tag{5.22}$$

$$Q_2(B,r) \approx \frac{(B-r+1)(B-r)}{2}\sum_{i=0}^{r} \frac{\pi(\langle i\rangle)}{B-i+1} + \sum_{i=B-r+1}^{b} i \sum_{j=0}^{B-i} \frac{\pi(\langle j\rangle)}{B-j+1}. \tag{5.23}$$

After calculating the quantities $\mathrm{CLP}_i(B,r)$ and $Q_i(B,r)$, the QoS metrics $\mathrm{CTD}_i(B,r)$, $i = 1, 2$ are calculated from formulas (5.5).

Note that the complexity of computing $CLP_i(B, r)$, $i = 1, 2$ is estimated as $O(Br)$ and that tabulated parameters $L(v, k)$ are used in the computation of these parameters.

Note 5.3 As it was indicated above, a complete sharing strategy is a special case of a partial sharing strategy. In fact, if we set $r = B$ in formulas (5.16)–(5.19) (or in formulas (5.20)–(5.23)), the corresponding results for CS scheme are obtained. So, in the case $r = B$, we have

$$CLP_1(B,B) = CLP_2(B,B) \approx \sum_{i=0}^{B} L(v_2, i)\pi(\langle B - i \rangle).$$

Thus, explicit (approximate) formulas to calculate the QoS metrics of the PS scheme have been obtained.

5.2.2 *Space Priorities Based on Push-Out Schemes*

Now consider the space priorities based on push-out scheme. For PO schemes, the newly arrived cell may substitute (i.e., push-out), in some cases, a cell of another type being already in the buffer. Here, we analyze PO schemes in which decisions about push-out are made only at the moment when new cell arrives if the buffer is full, i.e., a cell of any type is accepted at the buffer while there is a free space there. These schemes may be considered as a space priority mechanism, according to which priority cells push-out low-priority ones from the queue if there is no free space in the buffer when a high-priority cell arrives.

Let us consider the model of a switch which was described above (see Sect. 5.2.1). The proposed PO scheme of access (generalized push-out, GPO) is defined as follows. Arrived cell of type 1 (high-priority) pushes out cells of type 2 (low-priority) from the full buffer only when the current number of cells of type 1 is less than the given threshold c, $1 \le c \le B$; otherwise, i.e., when there are no cells of type 2 in the full buffer, arrived cell of type 1 is lost, and cells of type 2 are lost when the buffer is full. Note that when $c = B$, the given scheme reduces to the simple push-out scheme (SPO).

Note that this scheme might be treated as follows: the common buffer of size B is virtually divided between two kinds of traffic with preemption for cell of type 2, which gives a higher priority to cell of type 1 over cell of type 2. In other words, while buffer is not full it is used in accordance to complete sharing scheme, and when the buffer is full, arrived cell of type 1 will push out cell of type 2 if they (i.e., cells of type 2) already used their own limits (i.e., the number of cells of type 2 in buffer is more than $B - c$).

Operation of this switch can be described by a 2-D MC with the states of the form $\boldsymbol{n} = (n_1, n_2)$, where n_i specifies the number of packets of the ith type, $i = 1,2$. Then the state space of this chain is specified as follows:

$$S = \{\boldsymbol{n} : n_i = 0, 1, \ldots, B, i = 1,\ 2;\ n_1 + n_2 \le B\}. \tag{5.24}$$

The nonnegative elements of the Q-matrix of the given chain are determined as follows:

$$q(\boldsymbol{n}, \boldsymbol{n}') = \begin{cases} \lambda_i & \text{if } n_1 + n_2 < B,\ \boldsymbol{n}' = \boldsymbol{n} + \boldsymbol{e}_i,\ i = 1, 2, \\ \lambda_1 & \text{if } n_1 + n_2 = B,\ n_1 < c,\ n_2 > 0,\ \boldsymbol{n}' = \boldsymbol{n} + \boldsymbol{e}_1 - \boldsymbol{e}_2, \\ \mu_i & \text{if } \boldsymbol{n}' = \boldsymbol{n} - \boldsymbol{e}_i,\ i = 1, 2, \\ 0 & \text{in other cases.} \end{cases} \tag{5.25}$$

Let $S_{\mathrm{d}} = \{\boldsymbol{n} \in S : n_1 + n_2 = B\}$ denote the set of diagonal states. According to the introduced GPO scheme cell of type 1 is lost if upon its arrival system is in some state from the subset S_{d}^c of diagonal states, where $S_{\mathrm{d}}^c = \{\boldsymbol{n} \in S_{\mathrm{d}} : n_1 \ge c\}$. Then the loss probability of cells of type 1, $\mathrm{CLP}_1(B, c)$, is determined as follows:

$$\mathrm{CLP}_1(B, c) = \sum_{\boldsymbol{n} \in S_{\mathrm{d}}^c} p(\boldsymbol{n}). \tag{5.26}$$

Cells of type 2 are lost in the following cases: (a) upon arrival of this type of cell, system is in some of the diagonal states; (b) upon arrival of cell of type 1, system is in some of diagonal states of the kind $\boldsymbol{n} \in S_{\mathrm{d}} - S_{\mathrm{d}}^c$, i.e., in these cases an arrived cell of type 1 pushes out cell of type 2. Thus, $\mathrm{CLP}_2(B, c)$ is determined as follows:

$$\mathrm{CLP}_2(B, c) = \sum_{\boldsymbol{n} \in S_{\mathrm{d}}} p(\boldsymbol{n}) + \frac{v_{12}}{1 + v_{12}} \sum_{\boldsymbol{n} \in S_{\mathrm{d}} - S_{\mathrm{d}}^c} p(\boldsymbol{n}), \tag{5.27}$$

where $v_{12} = \lambda_1/\mu_2$.

The first term in formula (5.27) denotes the probability of event (a), while coefficient at the second term indicates the probability of the following event: during handling of cell of type 2 arrived at least one cell of type 1.

It is seen from Eqs. (5.26) and (5.27) that $\mathrm{CLP}_1(B, c) < \mathrm{CLP}_2(B, c)\ \forall c \in [1, B]$.

Cell transfer delays for cells of different types are calculated similar to Eq. (5.5) where the average number of cells of the ith type in the system, $Q_i(B, c)$, $i = 1, 2$, is determined by the stationary distribution of the initial model:

$$Q_i(B, c) = \sum_{k=1}^{B} k \xi_i(k), \tag{5.28}$$

where $\xi_i(k) = \sum_{\boldsymbol{n} \in S} {}^{p(\boldsymbol{n}) \delta(n_i, k)}$, $i = 1, 2$.

As in case space priorities based on PS scheme (see Sect. 5.2.1), here also there is no multiplicative form of stationary distribution of the investigated 2-D MC. Therefore, the only way to calculate QoS metrics under GPO scheme is to set up and solve an SGBE. For large size of buffer, this is a complex (even not solvable sometimes) computing problem; therefore, we propose here an approximate solution of the problem.

Let us consider the following splitting state space:

$$S = \cup_{i=0}^{B} S_i,\ S_i \cap S_j = \varnothing, \tag{5.29}$$

where $S_i = \{\mathbf{n} \in S : n_1 = i\},\ i = 0, 1, \ldots, B$.

As in Sect. 5.2.1, we will assume that $\lambda_2 >> \lambda_1$ and $\mu_2 >> \mu_1$. According to state space merging algorithm, classes of states S_i are united into individual merged states $\langle i \rangle$, and the appropriate merging function is introduced (see Eq. (5.8)). The merge function determines a merged model, which also is a 1-D BDP with the state space $\Omega = \{\langle i \rangle : i, = 0, 1, \ldots, B\}$.

Omitting intermediate mathematical calculations, let us present an algorithm for calculation of QoS metrics for $v_2 \neq 1$ (if at least one of the parameters v_1 or v_2 is equal to unity, then the formulas below can be simplified much more):

Step 1. Calculate

$$\pi(\langle i \rangle) = \begin{cases} v_1^i \pi(\langle 0 \rangle) & \text{if } 1 \le i \le c, \\ v_1^i \prod_{j=B-i+1}^{B-c} (1 - L(v_2, j)) \pi(\langle 0 \rangle) & \text{if } c + 1 \le i \le B, \end{cases} \tag{5.30}$$

where $\pi(\langle 0 \rangle)$ is determined from normalizing condition, i.e., $\sum_{i=0}^{B} \pi(\langle i \rangle) = 1$.

Step 2. Calculate

$$\mathrm{CLP}_1(B, c) \approx \sum_{i=c}^{B} L(v_2, B - i) \pi(\langle i \rangle); \tag{5.31}$$

$$\mathrm{CLP}_2(B, c) \approx \sum_{i=0}^{B} L(v_2, B - i) \pi(\langle i \rangle) + \frac{v_{12}}{1 + v_{12}} \sum_{i=0}^{c-1} L(v_2,\ B - i) \pi(\langle i \rangle); \tag{5.32}$$

$$Q_1(B, c) \approx \sum_{i=1}^{B} i \pi(\langle i \rangle); \tag{5.33}$$

$$Q_2(B, c) \approx \sum_{i=1}^{B} i \sum_{j=0}^{B-i} \rho_j(i) \pi(\langle\ \mathrm{j} \rangle). \tag{5.34}$$

After calculating the QoS metrics $\mathrm{CLP}_i(B, r)$ and $Q_i(B, r)$ from relations (5.31)–(5.34), the quantities $\mathrm{CTD}_i(B, r)$, $i = 1, 2$ are calculated from formulas (5.5).

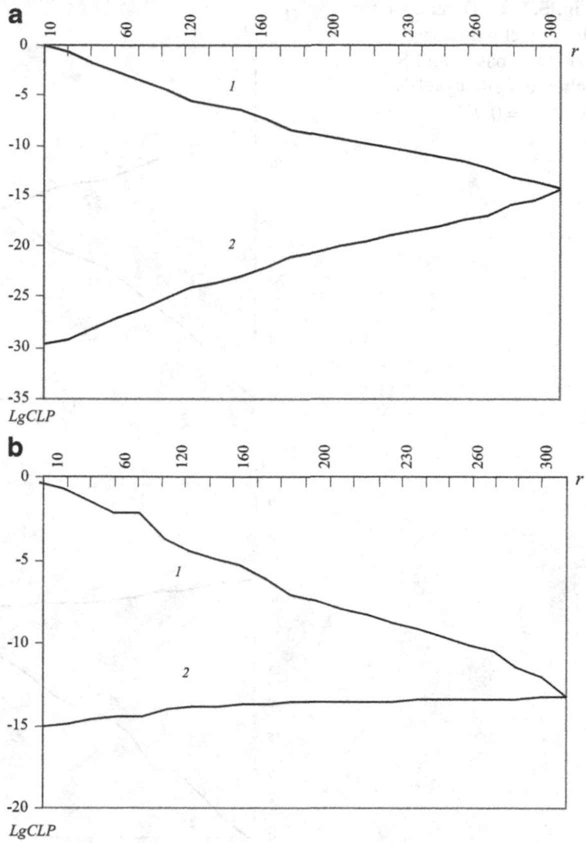

Fig. 5.1 CLP versus r for the model with space priorities based on PS scheme; (**a**)—$v_2 = 0.8$, (**b**)—$v_2 = 0.9$

In the special case $c = B$ from the given algorithm appropriate formulas for the SPO scheme are carried out.

5.2.3 *Numerical Results*

First consider some results of numerical experiments for the space priorities based on non-push-out scheme (i.e., SP based on PS scheme for buffer allocation).

Some of these results for symmetric ($v_1 = v_2$) and nonsymmetric ($v_1 \neq v_2$) loads with buffer size $B = 300$ are shown in Figs. 5.1 and 5.2. In order to be short, here only dependence of QoS metrics on threshold parameter r is shown.

The numerical results completely confirm the theoretical expectations concerning the behavior of the investigated QoS metrics $CLP_i(B, r)$ and $CTD_i(B, r)$, $i = 1, 2$. So, the increase in value of parameter r leads to increase in chances of cells of type 1 for acceptance in the common buffer, and thus, with growth of this parameter the probability of their loss decreases. At the same time, such increase simultaneously

Fig. 5.2 CTD versus r for the model with space priorities based on PS scheme; (**a**)—$v_2=0.8$, (**b**)—$v_2=0.9$

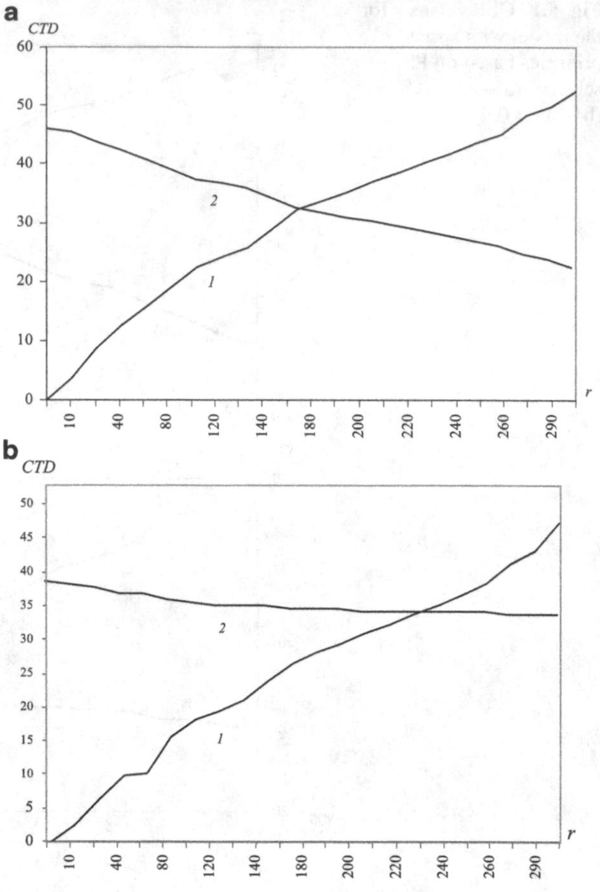

leads to reduction of chances of cells of type 2 for acceptance in the common buffer, i.e., the loss probability of cells of type 2 grows (see Fig. 5.1). From this figure we see that $\mathrm{CLP}_1(B,B)=\mathrm{CLP}_2(B,B)$.

As a result of analysis of the numerical results, we see that for fixed v_1, a slight decrease in v_2 leads to a substantial decrease in $\mathrm{CLP}_2(B,r)$, while if v_2 is decreased by 90 %, the decrease in $\mathrm{CLP}_2(B,r)$ becomes catastrophic. However, with fixed v_2, a decrease in v_1 does not have a strong effect on the value of $\mathrm{CLP}_i(B,r),\quad i=1,2$, particularly where the specified number of parameter r is small.

Increases in value of parameter r lead to decreasing of the loss probability of cells of type 1, i.e., the number of such cells in the buffer increases. Therefore average time of their waiting time in the buffer increases. However, conversely, the situation is observed at behavior research of an average waiting time in the buffer of cells of type 2 concerning parameter r (see Fig. 5.2).

Another goal of the computational experiments was to determine the accuracy of the proposed formulas. Note that the value of QoS metric determined by means of an SGBE was adopted as the exact one. Computational experiments to determine the exact values of QoS metrics were performed for small values of buffer size. Results of the comparison demonstrated that the proposed formulas were highly accurate, since in the worst case the difference between the exact and approximate values of QoS metrics did not exceed 0.1 %, this difference systematically decreasing as the size of the buffer grows (see Table 5.1).

On the other hand, by means of the proposed formulas the optimization problem relative to some (selected) performance criterion may be formulated and solved. As noted earlier, in real-world networks, packets of different types often impose different requirements on the quality of service. So, it is assumed that the QoS of different types of cells are estimated by means of their loss probabilities, i.e., constraints on the upper bounds of these QoS metrics are specified:

$$\mathrm{CLP}_i(B,r) \leq \varepsilon_i, \quad i = 1,\ 2, \tag{5.35}$$

where $\varepsilon_i > 0$, $i = 1, 2$ are the specified parameters (e.g., in ATM networks, ε_i varies within the range 10^{-10}–10^{-8}).

Here it will be assumed that the size of the switch buffer is fixed and that the load of the flows is uncontrolled. Then the optimization problem is formulated in the following way. It is required to find minimal ($\underline{r}$) and maximal ($\overline{r}$) values of parameter r for which constraints (5.35) are fulfilled, i.e., it is necessary to solve the following problem:

$$\overline{r} - \underline{r} \rightarrow \max \tag{5.36}$$

subject to Eq. (5.35).

In solving the latter problem the bounds for the $\mathrm{CLP}_i(B,r)$, $i = 1, 2$ will be important:

$$\mathrm{CLP}_1(B,B) \leq \mathrm{CLP}_1(B,r) \leq \mathrm{CLP}_1(B,1); \tag{5.37}$$

$$\mathrm{CLP}_2(B,1) \leq \mathrm{CLP}_2(B,r) \leq \mathrm{CLP}_2(B,B). \tag{5.38}$$

It is important to bear in mind that the ranges of variation (5.37) and (5.38) are un-improvable, since they are reached at certain combinations of parameters of the model.

In light of the above-indicated monotony properties of the investigated QoS metrics and unimprovable bounds (5.37) and (5.38), the following algorithm for solving problem (5.36) is proposed:

Step 1. If $\varepsilon_1 < \mathrm{CLP}_1(B,B)$ or $\varepsilon_2 < \mathrm{CLP}_2(B,1)$, the problem (5.36) has no solution.
Step 2. If $\varepsilon_1 > \mathrm{CLP}_1(B,1)$ and $\varepsilon_2 < \mathrm{CLP}_2(B,1)$, then $\underline{r} = 1$, $\overline{r} = B$.
Step 3. In parallel, the following problems are solved:

Table 5.1 Comparison for PS scheme, $B = 15, v_1 = 0.4\,\text{Erl}, v_2 = 0.8\,\text{Erl}$

	CLP_1		CLP_2		CTD_1		CTD_2	
r	EV	AV	EV	AV	EV	AV	EV	AV
1	0.809859	0.809967	0.007383	0.007477	0.422086	0.423581	0.035565	0.035666
2	0.654325	0.654455	0.007525	0.007673	0.430627	0.434031	0.035507	0.035643
3	0.527017	0.527265	0.007665	0.00769	0.459989	0.462735	0.035452	0.035503
4	0.422788	0.4238	0.007799	0.007814	0.487749	0.491038	0.035399	0.035497
5	0.337469	0.339511	0.007928	0.008044	0.513641	0.533078	0.035349	0.035405
6	0.267664	0.268015	0.008049	0.008235	0.537478	0.539841	0.035303	0.035398
7	0.210603	0.211535	0.008162	0.008455	0.55915	0.560673	0.035261	0.035377
8	0.164009	0.167103	0.008266	0.008573	0.578619	0.581105	0.035224	0.035301
9	0.126014	0.126134	0.00836	0.008701	0.595914	0.601749	0.03519	0.035215
10	0.095077	0.096175	0.008445	0.008799	0.611114	0.614067	0.03516	0.035217
11	0.069924	0.079167	0.00852	0.008801	0.624343	0.628106	0.035134	0.035204
12	0.049506	0.050734	0.008586	0.008898	0.635752	0.639058	0.035112	0.035218
13	0.032957	0.033106	0.008645	0.008899	0.645511	0.648058	0.035093	0.035198
14	0.019564	0.020174	0.008695	0.0089	0.653798	0.658904	0.035077	0.035147
15	0.008739	0.009015	0.008739	0.008937	0.660791	0.663065	0.035063	0.035106

Table 5.2 Results of a solution of problem (5.28) in the case $B = 100$, $v_1 = 0.9$ Erl, $\varepsilon_1 = 10^{-2}$

v_2	ε_2	$[\underline{r}, \overline{r}]$
0.2	10^{-5}	$\varnothing$
0.2	10^{-6}	[23,99]
0.2	10^{-7}	[23,97]
0.2	10^{-8}	[23,96]
0.2	10^{-5}	[23,94]
0.2	10^{-10}	[23,93]
0.8	10^{-5}	$\varnothing$
0.8	10^{-6}	[35,93]
0.8	10^{-7}	[35,75]
0.8	10^{-8}	[35,55]
0.8	10^{-5}	[35,35]
0.8	10^{-10}	$\varnothing$

$$r_1 = \arg\min_r \{\mathrm{CLP}_1(B, r) \leq \varepsilon_1\}.$$

$$r_2 = \arg\max_r \{\mathrm{CLP}_2(B, r) \leq \varepsilon_2\}.$$

The dichotomy method may be used for solution of the last problems.

Step 4. If $r_1 > r_2$, then problem (5.36) has no solution; otherwise its solution will be $\underline{r} = r_1, \overline{r} = r_2$.

The results of computational experiments for the solution of problem (5.36) are presented in Table 5.2. Hereinafter, the symbol $\varnothing$ denotes that the problem does not have a solution.

Analyses of these data lead us to the following conclusions:

For fixed loads of flows, as ε_2 decreases, $\overline{r}$ also decreases, while $\underline{r}$ remains unchanged.

For fixed ε_2, as v_2 grows, $\underline{r}$ increases, while $\overline{r}$ decreases.

It is evident that CS scheme of buffer sharing is an absolutely fair servicing, in the sense of equality of the loss probabilities for different types of cells; it is realized in PS scheme when $r = B$. In practice, however, sometimes such an absolutely fair servicing is not required, and then the concept of an ε-fair servicing may be introduced. In ε-fair servicing the difference between the loss probabilities of different types of cells does not exceed some specified $\varepsilon > 0$. So, the problem of finding an ε-fair servicing with the use of the PS scheme of buffer sharing is formulated mathematically as follows:

$$r^* = \arg\min_r \{\mathrm{CLP}_1(B, r) - \mathrm{CLP}_2(B, r) < \varepsilon\}. \quad (5.39)$$

By taking into account monotony properties of the functions $\mathrm{CLP}_i(B, r)$,

Table 5.3 Results of a solution of problem (5.31) in the case $B = 100$, $\nu_2 = 0.8$ Erl

ν_1	ε	r^*
0.9	10^{-2}	35
0.9	10^{-4}	78
0.9	10^{-5}	97
0.9	10^{-6}	100
0.1	10^{-4}	42
0.1	10^{-5}	52
0.1	10^{-6}	63
0.1	10^{-7}	73
0.1	10^{-8}	83
0.1	10^{-5}	93
0.1	10^{-10}	99
0.1	10^{-11}	100

$r = 1, 2$, the method of dichotomy may be used to solve problem (5.39). The initial interval of uncertainty $[1, B]$ is divided into half, i.e., a point $b = \text{int}((B+1)/2)$ is found, where int(x) denotes the integer part of x. If the condition of problem (5.39) is satisfied at this point, the interval $[1, b]$ is considered next; otherwise, the interval $[b, B]$ is considered. The condition that determines the termination of the algorithm is as follows: find an interval of unit length such that the right endpoint satisfies the condition of problem (5.39) but the left endpoint does not satisfy it. It is precisely this right endpoint which is the desired quantity r^*. The algorithm terminates in a finite number of steps; in the worst case the condition of problem (5.39) will be satisfied at the point $r = B$.

The results of a solution of problem (5.39) are shown in Table 5.3. As it was expected, r^* grows with decreasing ε.

Now consider some results of numerical experiments for the space priorities based on push-out scheme. Formulas found in Sect. 5.2.2 allow to analyze the performance of SP based on GPO scheme of access. For brevity here we omit the detailed analysis of the corresponding graphs, and below we give short comments related to performance of the GPO scheme.

Numerical experiments show that function $\text{CLP}_1(B, c)$ is decreasing one while function $\text{CLP}_2(B, c)$ is increasing one versus threshold c subject to fixed loads and buffer size. It is interesting to note that both functions $\text{CLP}_1(B, c)$ and $\text{CLP}_2(B, c)$ are increasing one with respect to loads of traffics and parameter ν_{12}. Function $Q_1(B, c)$ is increasing one, while function $Q_2(B, c)$ is decreasing one with respect to threshold parameter c. At the same time, the behavior of both functions $\text{CTD}_1(B, c)$ and $\text{CTD}_2(B, c)$ essentially depends on the rate of change of functions $\text{CLP}_i(B, c)$ and $Q_i(B, c)$, $i = 1, 2$, i.e., their monotonic properties for any initial data are not guaranteed.

i. At the same time, the behavior of the QoS metrics versus loading parameters of model (at fixed values of B and c) is quite predicted. For instance, the function $\text{CTD}_2(B, c)$ is nonincreasing one with respect to handling intensity of cell of type

Table 5.4 Comparison for QoS metrics of cells of the first type in GPO scheme

c	CLP_1		CTD_1	
	EV	AV	EV	AV
5	4.79E-02	4.18E-02	0.704096	0.711035
10	1.02E-02	1.00E-02	0.887377	0.882259
15	2.38E-03	2.11E-03	0.962189	0.977513
20	5.60E-04	5.03E-04	0.988287	0.990452
25	1.32E-04	1.07E-04	0.996561	0.997430
30	3.15E-05	2.29E-05	0.999026	0.999157
35	7.48E-06	6.99E-06	0.999731	0.999789
40	1.78E-06	1.05E-06	0.999927	0.999871
45	4.43E-07	2.46E-07	0.999979	0.999980
49	1.71E-07	8.02E-08	0.999992	0.999997

Table 5.5 Comparison for QoS metrics of cells of the second type in GPO scheme

c	CLP_2		CTD_2	
	EV	AV	EV	AV
5	0.702105	0.699005	19.437348	20.068437
10	0.739757	0.723128	21.949898	21.999958
15	0.747625	0.739056	22.512933	22.668755
20	0.749439	0.748854	22.634315	22.788849
25	0.749867	0.748910	22.659983	22.789900
30	0.749685	0.748943	22.665318	22.790341
35	0.749993	0.748968	22.666406	22.794153
40	0.749998	0.748978	22.666626	22.795037
45	0.749999	0.749788	22.666679	22.795102
49	0.750000	0.749997	22.666708	22.797305

2 (i.e., μ_2). It is necessary to notice that with growth of the parameter μ_2, the function $\mathrm{CLP}_1(B,c)$ decreases and the function $Q_1(B,c)$ increases. These facts are explained as follows: with the growth of parameter μ_2 the speed of clearing of the buffer from cells of type 2 is increased and that chance of cells of type 1 to capture the place in buffers increases. Thus for function $Q_1(B,c)$ such dependence is especially brightly observed at small values of the threshold parameter c since at large values of the specified parameter, arriving cells of type 1 at the overflowed buffer almost always supersede cells of the type 2.

Here we show only high accuracy of the proposed formulas for SP based on GPO scheme (see Tables 5.4 and 5.5) where $B=50, v_1=0.75, v_2=1.6, v_{12}=0.6$.

Optimization problem for SP based on GPO scheme for the determination of the minimal ($\underline{c}$) and maximal ($\overline{c}$) values of parameter c for which constraints (5.30) are fulfilled is set as follows: it is necessary to solve the following problem:

$$\overline{c}-\underline{c}\rightarrow \max \tag{5.40}$$

subject to Eq. (5.35).

Table 5.6 Results of a solution of problem (5.40) in the case $B = 100$

λ_1	μ_1	λ_2	μ_2	ε_1	ε_2	$[\underline{c}, \overline{c}]$
0.2	0.5	5	5	10^{-1}	10^{-5}	Ø
0.2	0.5	5	6	10^{-2}	10^{-5}	[1, 98]
0.2	0.5	5	12	10^{-2}	10^{-5}	[1, 98]
0.2	0.5	10	6	10^{-4}	10^{-5}	Ø
0.3	0.5	5	6	10^{-8}	10^{-5}	[1, 98]
0.3	0.5	5	6	10^{-9}	10^{-5}	[4, 98]
0.3	0.5	5	6	10^{-11}	10^{-7}	[11, 98]
0.3	0.5	5	2	10^{-6}	10^{-1}	[25, 90]
0.4	0.5	5	2	10^{-6}	10^{-1}	[55, 90]
0.5	0.5	5	2	10^{-6}	10^{-1}	Ø
0.5	0.8	5	2	10^{-7}	10^{-2}	[16, 98]
0.5	0.8	5	2	10^{-6}	10^{-2}	[14, 98]
0.5	0.9	5	2	10^{-6}	10^{-2}	[12, 98]
0.5	0.9	5	2	10^{-8}	10^{-1}	[17, 98]
0.3	0.5	5	10	10^{-8}	10^{-6}	[1, 98]
0.3	0.5	5	12	10^{-8}	10^{-6}	[1, 98]
0.3	2	5	12	10^{-8}	10^{-6}	[1, 98]
0.3	2	5	12	10^{-6}	10^{-6}	Ø

Before starting to describe the algorithm of solution of problem (5.40), note that similar to Eqs. (5.37) and (5.38) the unimprovable bounds for these QoS metrics might be obtained. This allows us to propose the following algorithm for solution of problem (5.40):

Step 1. If $\text{CLP}_1(B,B) > \varepsilon_1$ and $\text{CLP}_2(B,1) > \varepsilon_2$, then problem (5.40) has no solution.
Step 2. If $\text{CLP}_1(B,1) \le \varepsilon_1$ and $\text{CLP}_2(B,B) \le \varepsilon_2$, then $\underline{c} = 1, \overline{c} = B$.
Step 3. In parallel, the following problems are solved:

$$c_1 = \arg\min_c \{\text{CLP}_1(B,c) \le \varepsilon_1\};$$

$$c_2 = \arg\max_c \{\text{CLP}_2(B,c) \le \varepsilon_2\}.$$

Step 4. If $c_1 > c_2$ then problem (5.40) has no solution. Otherwise, the solution of problem (5.35) is $\underline{c} = c_1, \overline{c} = B$.

The results of solution of problem (5.40) for a sample model are shown in Table 5.6. As in the given problem, if there are a large number of initial data, it is difficult to draw the general conclusions for its optimal solution.

Similar to problem (5.39) finding of an ε-fair servicing for space priorities based on GPO scheme may be formulated and solved. We left it to the reader.

5.3 Models of Switches with Common Output Ports

In this section we are considering combination of the space and time priorities in buffer allocation schemes for the switch in which the outgoing ports are common for both types of packets. Here we classify multimedia traffics into two types of traffic: traffic of type 1 (such as data) is delay insensitive and loss sensitive, and traffic of type 2 (such as voice) is delay sensitive and loss insensitive. As it was mentioned above, application of fixed priority scheme for one type of traffic (either for type 1 or type 2) in such multimedia traffic is unacceptable since lowerpriority traffic will have both large delay and high loss which does not satisfy the requirements of heterogeneous packets. This is why in switch with common output ports we need multiple priorities such as both high TP for delay-sensitive traffic and high SP for loss-sensitive traffic.

The multiple priorities might be realized by using different schemes. In other words, SP might use both non-push-out and push-out schemes for loss-sensitive traffic, and TP might use both preemptive and non-preemptive priority schemes for delay-sensitive traffic. As it was mentioned in Sect. 5.1, the scheduler must be easy to implement in hardware. According to this recommendation here we assume that time priorities are based on non-preemptive scheme. At the same time, proposed here is the approach that allows to investigate the TP based on preemptive priorities. However, we consider both SP based on push-out and non-push-out schemes.

5.3.1 Multiple Priorities Based on Non-Push-Out Scheme

A single output port of switch is shared by common finite buffer of size B for two types of cells; cells of type i arrive according to a Poisson process with intensity λ_i, $i=1,2$ and are serviced by exponentially distributed time with the same (common) parameter μ. We assume that cells of type 1 are delay insensitive and loss sensitive and cells of type 2 are delay sensitive and loss insensitive. High TP for delay-sensitive traffic and high preventive SP based on non-push-out scheme for loss-sensitive traffic are applied. In other words, the switch drops arriving cell when the queue size of cells of type 2 exceeds a certain threshold.

Since the cells of type 1 are more sensitive to possible loss due to buffer overflow than cells of type 2, cells of type 1 have higher SP based on non-push-out scheme with a threshold r as follows. Cells of type 2 are received only when at the time they arrive, the number of cells of the given type is less than some specified threshold r, $0<r\leq B$; otherwise these cells are lost. Cells of type 1 will be lost if and only if the buffer is full at the time of arrival, i.e., they are always received into the buffer if there is at least a single free space in the buffer. However, high non-preemptive TP for delay-sensitive traffic (i.e., for traffic of type 2) is applied as follows: whenever channel is available for traffic, output port chooses the cell of type 2, and the cell of type 1 is chosen only when there is no cell of type 2 in the queue.

Mathematical model of the given system is the following 2-D MC. The states of the indicated Markov chain have the form $\boldsymbol{n} = (n_1, n_2)$ where n_i is the number of cells of type i in the system, $i = 1, 2$. So, state space of this 2-D MC is

$$S = \{\boldsymbol{n} : n_1 = 0, 1, \ldots, B;\ n_2 = 0, 1, \ldots, r;\ n_1 + n_2 \le B\}.$$

Nonnegative elements of the given 2-D MC are determined as follows:

$$q(\boldsymbol{n}, \boldsymbol{n}') = \begin{cases} \lambda_1 & \text{if } n_1 + n_2 < B,\ \boldsymbol{n}' = \boldsymbol{n} + \boldsymbol{e}_i, \\ \lambda_2 & \text{if } n_2 < r,\ \boldsymbol{n}' = \boldsymbol{n} + \boldsymbol{e}_2, \\ \mu & \text{if } n_2 = 0,\ \boldsymbol{n}' = \boldsymbol{n} - \boldsymbol{e}_1 \text{ or } n_2 > 0,\ \boldsymbol{n}' = \boldsymbol{n} - \boldsymbol{e}_2, \\ 0 & \text{in other cases.} \end{cases} \tag{5.41}$$

The given 2-D MC always has a stationary distribution since the state space is finite and all its states communicate. The exact analysis of the QoS metrics of the investigated system is based on solving an appropriate GSBE which is constructed by using relations (5.41) (we left it to the reader). To overcome the known computational difficulties of exact analysis below we develop simple (approximate) explicit formulas.

The loss probability for cells of type i, $i = 1, 2$ are determined as

$$\text{CLP}_1(B, r) = \sum_{i=B-r+1}^{B} p(i,\ B - i);$$

$$\text{CLP}_2(B, r) = \sum_{i=0}^{B-r} p(i, r) + \sum_{i=B-r+1}^{B} p(i,\ B - i).$$

Average number of cells of type i, $i = 1, 2$, in the buffer ($Q_i(B, r)$), is obtained similar to Eq. (5.6), and their average waiting time ($\text{CTD}_i(B, r)$) is calculated similar to Eq. (5.5).

Approximate analysis of the given 2-D MC is based on splitting its state space similar to Eq. (5.29). Next, the class S_i in splitting Eq. (5.29) is merged into one state $\langle i \rangle$, and merged space $\Omega = \{\langle i \rangle : 0 \le i \le B\}$ is defined. Steady-state probabilities within class S_i are calculated as follows (for brevity, we consider here the case $v_2 \neq 1$ where $v_2 = \lambda_2/\mu$):

Case $0 \le i \le B - r$:

$$\rho_i(j) = \frac{v_2^j (1 - v_2)}{1 - v_2^{r+1}}, \quad j = 0, 1, \ldots, r;$$

Case $B - r + 1 \le i \le B$:

$$\rho_i(j) = \frac{v_2^j(1-v_2)}{1-v_2^{B+1-i}}, \quad j = 0, 1, \ldots, B-i.$$

By using the last formulas, we obtain the following relations to calculate the elements of Q-matrix of the merged model:

$$q(i,j) = \begin{cases} \lambda_1 & \text{if } 0 \le i \le B-r-1, j=i+1, \\ \lambda_1(1-L(v_2, B-i)) & \text{if } B-r \le i < B, j=i+1, \\ \mu\rho_i(0) & \text{if } j=i-1, \\ 0 & \text{in other cases.} \end{cases}$$

Therefore, the stationary distribution of the merged model is

$$\pi(\langle i \rangle) = \begin{cases} v_1^i\left(\prod_{j=1}^{i}\rho_j(0)\right)^{-1}\pi(\langle 0 \rangle) & \text{if } 1 \le i \le B-r, \\ v_1^i\left(\prod_{j=1}^{i}\rho_j(0)\right)^{-1}\prod_{j=B-i+1}^{r}(1-L(v_2,j))\pi(\langle 0 \rangle) & \text{if } B-r+1 \le i \le B, \end{cases}$$

where $\pi(\langle 0 \rangle)$ is obtained from normalizing condition.

After performing the necessary mathematical transformations, one obtains the following approximate formulas for the cell loss probabilities:

$$CLP_1(B,r) \approx \sum_{i=B-r+1}^{B} L(v_2, B-i)\pi(\langle i \rangle); \tag{5.42}$$

$$CLP_2(B,r) \approx L(v_2, r)\sum_{i=0}^{B-r}\pi(\langle i \rangle) + CLP_1(B,r); \tag{5.43}$$

Approximate values of QoS metrics $Q_i(B,r), i = 1,2$ are obtained from formulas (5.33) and (5.34), respectively (in Eq. (5.34), the upper limit of the first sum must be substituted by parameter r). Cell transfer delays for heterogeneous cells are calculated by using formula (5.5).

5.3.2 *Multiple Priorities Based on Push-Out Scheme*

Now consider model of switch in which SP are based on push-out scheme. So, fixed non-preemptive TP and reactive SP based on push-out scheme are applied in switch. In other words, reactive SP decide to drop the packet during period of congestion, i.e., when buffer of switch has become full and new packet has arrived. The detailed description of model consists in the following.

Traffic of type 1 has higher SP based on push-out scheme with a threshold c as follows. Cell of type 1 will be lost if and only if the buffer is full at the time of its arrival and the number of cell of type 1 in the buffer is greater than or equal to c, $1 \leq c \leq B$, and if the buffer is full at the time of arrival of type 1 cell and the number of such cell in the buffer is less than c, then this cell pushes out a cell of type 2 from the buffer and takes its place. An arriving cell of type 2 will be lost if at the arrival moment the buffer is full.

As in previous case, since traffic of type 2 is more sensitive to possible delay than traffics of type 1, traffic of type 2 has higher strict non-preemptive priority, i.e., whenever channel is available for traffic, server chooses the traffic of type 2, and the traffic of type 1 is chosen only when there is no traffic of type 2 in the queue.

The exact analysis of the given model is based on using GSBE of appropriate 2-D Markov chain. The states of the mentioned chain have the form $\boldsymbol{n} = (n_1, n_2)$ where n_i is the number of cells of type i in the system, $i = 1, 2$. Thus, state space of the 2-D MC is defined as Eq. (5.24), while its transition rates are determined as follows:

$$q(\boldsymbol{n}, \boldsymbol{n}') = \begin{cases} \lambda_i & \text{if}\, n_1 + n_2 < B,\ \boldsymbol{n}' = \boldsymbol{n} + \boldsymbol{e}_i,\ i = 1, 2, \\ \lambda_1 & \text{if}\, n_1 + n_2 = B,\ n_1 < c,\ \boldsymbol{n}' = \boldsymbol{n} + \boldsymbol{e}_1 - \boldsymbol{e}_2, \\ \mu & \text{if}\, n_2 = 0,\ \boldsymbol{n}' = \boldsymbol{n} - \boldsymbol{e}_1\, \text{or}\, n_2 > 0,\ \boldsymbol{n}' = \boldsymbol{n} - \boldsymbol{e}_2, \\ 0 & \text{in other cases.} \end{cases} \tag{5.44}$$

The given 2-D MC always has a stationary distribution since the state space is finite and all its states communicate. Its stationary distribution is obtained by solving GSBE which is constructed by using relations (5.44) (we left it to reader).

Then the loss probability for cells of types 1 and 2 is determined from formulas (5.26) and (5.27), respectively, where in Eq. (5.27) coefficient v_{12} is substituted by v_1.

The average number of cells of type i, $i = 1, 2$ in the buffer is obtained by Eq. (5.28), and their average waiting time is calculated similar to Eq. (5.5).

Approximate analysis of the given 2-D MC is based on splitting its state space similar to Eq. (5.29). Omitting intermediate mathematical calculations, let us present an algorithm for calculation of QoS metrics for $v_2 \neq 1$. So, the conditional stationary distribution within class S_i coincides with steady-state probabilities of the queueing model $M(\lambda_2)/M(\mu)/1/B - i$:

$$\rho_i(j) = \frac{v_2^j(1 - v_2)}{1 - v_2^{B+1-i}}, \quad j = 0, 1, \ldots, B - i.$$

So, by using Eq. (5.46), we define Q-matrix of the merged model:

$$q(i,j) = \begin{cases} \lambda_1 & \text{if } 0 \le i \le c, j = i+1, \\ \lambda_1(1 - L(v_2, i)) & \text{if } c+1 \le i < B, j = i+1, \\ \mu\rho_i(0) & \text{if } j = i-1, \\ 0 & \text{in other cases.} \end{cases}$$

Thus the stationary distribution of the merged model is obtained as follows:

$$\pi(\langle i \rangle) = \begin{cases} v_1^i \left(\prod_{j=1}^{i} \rho_j(0)\right)^{-1} \pi(\langle 0 \rangle) & \text{if } 1 \le i \le c, \\ v_1^i \left(\prod_{j=1}^{i} \rho_j(0)\right)^{-1} \prod_{j=B-i+1}^{B-c} (1 - L(v_2, j))\pi(\langle 0 \rangle) & \text{if } c+1 \le i \le B, \end{cases}$$

where $\pi(\langle 0 \rangle)$ is obtained from normalizing condition.

Finally, we find the following approximate formula for the QoS metrics:

$$\mathrm{CLP}_1(B,c) \approx \sum_{i=c+1}^{B} L(v_2,\, B-i)\pi(\langle i \rangle); \tag{5.45}$$

$$\mathrm{CLP}_2(B,c) \approx \sum_{i=0}^{B} L(v_2, B-i)\pi(\langle i \rangle) + \frac{v_1}{1+v_1} \sum_{i=0}^{c-1} L(v_2,\, B-i)\pi(\langle i \rangle). \tag{5.46}$$

Approximate values of QoS metrics $Q_i(B,r)$, $i = 1, 2$ are obtained exactly from formulas (5.33) and (5.34), respectively. Cell transfer delays for heterogeneous cells are from Eq. (5.5). So, calculation of QoS metrics of the model with multiple priorities is obtained by simple computational procedures.

5.3.3 Numerical Results

First consider results of numerical experiments for multiple priorities based on non-push-out scheme. In order to be short here only the dependency of QoS metrics on the threshold parameter r is shown in Figs. 5.3 and 5.4 where buffer size and loads are fixed, i.e., in numerical examples the initial data are selected as follows: $B = 30, \lambda_1 = 0.3, \lambda_2 = 5, \mu = 2$.

With the growth of parameter r, chances for acceptance in the buffer of cells of type 2 grow, and thus, the probability of their loss decreases and at the same time the loss probability of cells of type 1 increases (see Fig. 5.3). Note that the rate of change of functions $\mathrm{CLP}_k(B,r)$, $k = 1, 2$ is very slow especially for large values of parameter r and for $r \ge 7$ both functions become almost constant, and moreover we have $\mathrm{CLP}_1(B,r) \approx \mathrm{CLP}_2(B,r)$. In other words, for some values of threshold parameter r, loss probabilities of loss-sensitive and loss-tolerance traffics become the same. However, with growth of the given parameter, both functions $\mathrm{CTD}_1(B,r)$ and $\mathrm{CTD}_2(B, r)$ increase (see Fig. 5.4). The last facts are explained as follows: with

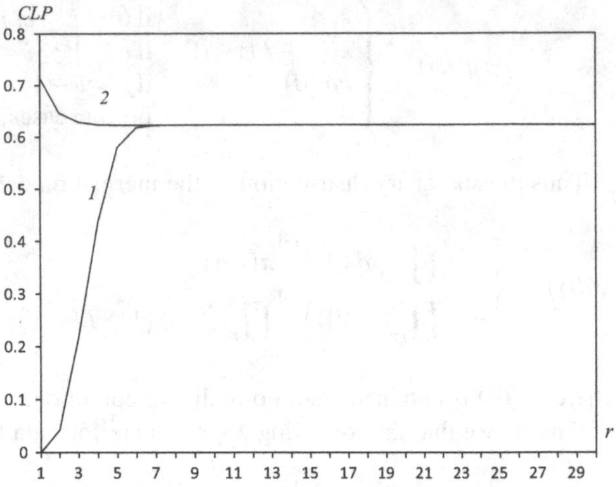

Fig. 5.3 CLP versus r for the model with multiple priorities based on NPO scheme; 1—CLP_1, 2—CLP_2

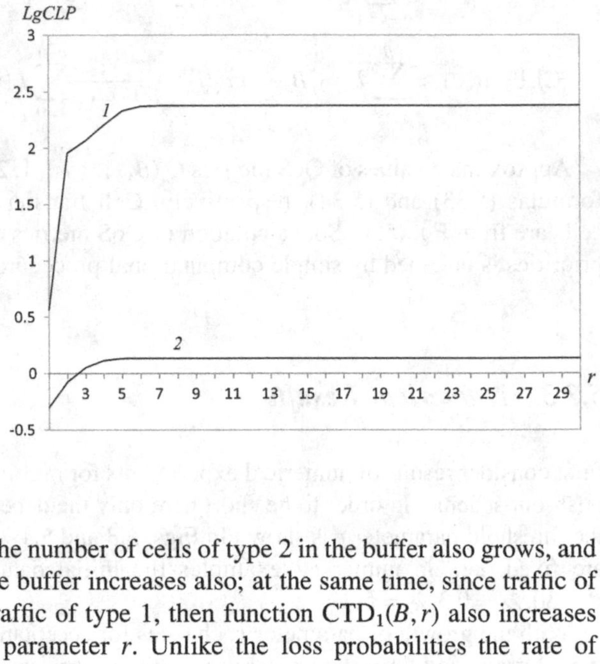

Fig. 5.4 CTD versus r for the model with multiple priorities based on NPO scheme; 1—CTD_1, 2—CTD_2

increase in this parameter the number of cells of type 2 in the buffer also grows, and as a result their delay in the buffer increases also; at the same time, since traffic of type 2 has high TP over traffic of type 1, then function $CTD_1(B,r)$ also increases with respect to threshold parameter r. Unlike the loss probabilities the rate of changes of functions $CTD_k(B,r)$, $k=1,2$ is very high at small values of parameter r; however, again for $r \geq 7$ this function becomes almost constant. At the same time note that $CTD_2(B,r) << CTD_1(B,r)$ for any values of parameter r, i.e., the proposed multiple priority scheme allows to fulfill the required QoS level for delay-sensitive traffic.

Now consider the results of numerical experiments for multiple priorities based on push-out scheme. In order to be short here only the dependency of QoS metrics on the threshold parameter c is shown in Figs. 5.5 and 5.6 where buffer size and

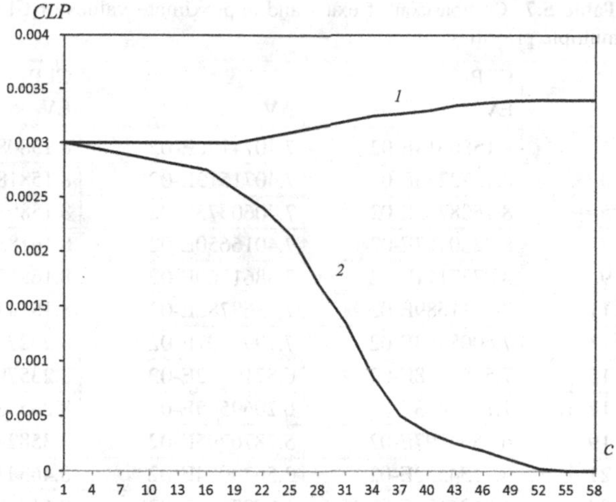

Fig. 5.5 CLP versus c for the model with multiple priorities based on PO scheme; 1—CLP_1, 2—CLP_2

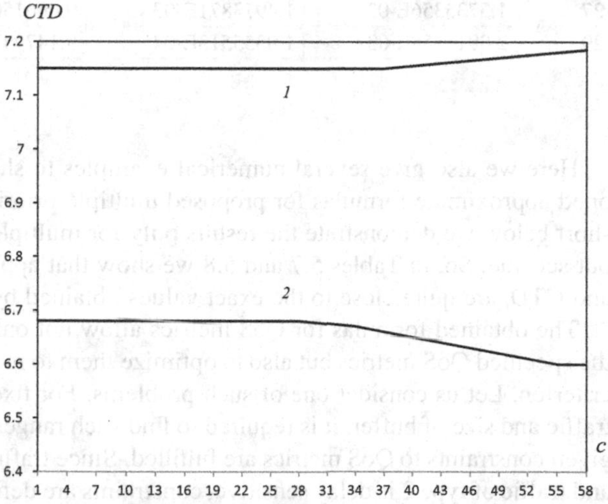

Fig. 5.6 CTD versus c for the model with multiple priorities based on PO scheme; 1—CTD_1, 2—CTD_2

loads are fixed. In these numerical examples, we select the following initial data: $B = 30, \lambda_1 = 0.01, \lambda_2 = 0.08, \mu = 1/12$.

As it was expected, with growth of parameter c chances for acceptance in the buffer of cells of type 1 grow, and thus, the probability of their loss decreases and at the same time the loss probability of cells of type 2 increases (see Fig. 5.5). Note that with growth of the given parameter function $CTD_1(B, c)$ increases, and $CTD_2(B, c)$ decreases (see Fig. 5.6). The last facts are explained as follows: with increase in this parameter the number of cells of type 1 in the buffer also grows, and simultaneously as a result of push-out from the buffer, the number of cells of type 2 in the buffer decreases.

Table 5.7 Comparison of exact and approximate values of CLP for the model of switch with multiple priorities

	CLP_1		CLP_2	
c	EV	AV	EV	AV
1	8.1580614E-02	7.4074126E-02	8.1580844E-02	7.4074757E-02
3	8.1572253E-02	7.4071513E-02	8.1581889E-02	7.4075058E-02
5	8.1508712E-02	7.4060373E-02	8.1589832E-02	7.4076423E-02
7	8.1280457E-02	7.4016650E-02	8.1618364E-02	7.4081858E-02
9	8.0737111E-02	7.3861260E-02	8.1686282E-02	7.4101248E-02
11	7.9711689E-02	7.3368782E-02	8.1814460E-02	7.4162773E-02
13	7.8008493E-02	7.1999697E-02	8.2027359E-02	7.4333872E-02
15	7.5363698E-02	6.8719692E-02	8.2357959E-02	7.4743835E-02
17	7.1396643E-02	6.2069565E-02	8.2853841E-02	7.5575066E-02
19	6.5564797E-02	5.0876795E-02	8.3582821E-02	7.6974133E-02
21	5.7154532E-02	3.5629074E-02	8.4634104E-02	7.8880076E-02
23	4.5421671E-02	1.9523734E-02	8.6100712E-02	8.0893231E-02
25	3.0217959E-02	7.3169717E-03	8.8001176E-02	8.2419071E-02
27	1.3735356E-02	1.4973871E-03	9.0061501E-02	8.3146518E-02
29	2.3915558E-03	1.0355513E-04	9.1479477E-02	8.3320747E-02

Here we also give several numerical examples to show the accuracy of developed approximate formulas for proposed multiple priority schemes. In order to be short below we demonstrate the results only for multiple priorities based on push-out scheme. So, in Tables 5.7 and 5.8 we show that approximation values of CLP_i and CTD_i are quite close to the exact values obtained by the balanced equations.

The obtained formulas for QoS metrics allow not only to study the behavior of the specified QoS metrics but also to optimize them to meet a certain service quality criterion. Let us consider one of such problems. For fixed values of parameters of traffic and size of buffer, it is required to find such ranges of value within which the given constraints to QoS metrics are fulfilled. Since traffic of type 1 is loss sensitive and traffic of type 2 is delay sensitive, constraints are defined only for these kinds of QoS metrics. This suggests solving the following problem:

$$\overline{c} - \underline{c} \to \max \tag{5.47}$$

s.t. $CLP_1(B,c) \le \varepsilon_1$ and $CTD_2(B,c) \le \varepsilon_2$ $\forall c \in [\underline{c}, \overline{c}]$.

Some numerical results are given in Table 5.9.

Here we consider the sensitivity of optimal solution with respect to the change of traffic load because the traffic rate may be varied while systems are working. The optimal solutions of the problem in Eq. (5.47) are little affected by the change of λ_2. For instance, at $B=70; \mu=1; \lambda_1=0.4; \lambda_2=6.3$ for the given $\varepsilon_1=2.5E-0.8; \varepsilon_2=10$, the optimal solution is [67, 69], and this interval is changed as follows:

Table 5.8 Comparison of exact and approximate values of CTD for the model of switch with multiple priorities

	CTD_1		CTD_2	
c	EV	AV	AV	EV
1	1.6913177E+03	2.0440386E+03	6.9524129E+01	6.7557307E+01
3	1.6913191E+03	2.0440427E+03	6.9522098E+01	6.7556877E+01
5	1.6913189E+03	2.0440597E+03	6.9508105E+01	6.7555040E+01
7	1.6912763E+03	2.0441159E+03	6.9463046E+01	6.7548219E+01
9	1.6910765E+03	2.0442700E+03	6.9368094E+01	6.7525738E+01
11	1.6905175E+03	2.0446084E+03	6.9211745E+01	6.7460540E+01
13	1.6892959E+03	2.0451283E+03	6.8989089E+01	6.7296880E+01
15	1.6869640E+03	2.0453659E+03	6.8699049E+01	6.6948435E+01
17	1.6828537E+03	2.0438275E+03	6.8344293E+01	6.6333216E+01
19	1.6759882E+03	2.0379494E+03	6.7935066E+01	6.5455459E+01
21	1.6650468E+03	2.0257374E+03	6.7496977E+01	6.4479103E+01
23	1.6485941E+03	2.0087581E+03	6.7081383E+01	6.3682756E+01
25	1.6261442E+03	1.9930056E+03	6.6769804E+01	6.3258212E+01
27	1.6010313E+03	1.9841532E+03	6.6641409E+01	6.3140980E+01
29	1.5834234E+03	1.9817098E+03	6.6663420E+01	6.3133124E+01

Table 5.9 Solution results for the problem (5.47)

B	50	70	70	70	70	50	50
λ_1	0.4	0.4	0.4	0.4	0.4	0.4	0.4
λ_2	0.5	0.5	0.5	0.5	6.3	0.5	0.5
ε_1	2E-06	2E-08	2.5E-08	2.56E-08	2.56E-08	2.56E-08	2E-06
ε_2	2	5	5	10	10	10	10
$[\underline{c}, \overline{c}]$	[46, 49]	[68, 69]	[63, 69]	[63, 69]	[57, 69]	[67, 69]	Ø

$$[\underline{c}, \overline{c}] = \begin{cases} [68, 69] & \text{if } \lambda_2 \in [6.4, 7.3], \\ [69, 69] & \text{if } \lambda_2 \in [7.4, 19.5], \\ \varnothing & \text{if } \lambda_2 > 19.5. \end{cases}$$

So we may conclude that the optimal solutions of the given problem (5.47) are almost insensitive with respect to the arrival rate of traffics of type 2. Analogous results were observed with respect to the arrival rate of traffics of type 1.

At the end of this section we give the computation time ratio of exact and approximation analysis (see Table 5.10). From this table it is concluded that with growth of the buffer size, the efficiency of the approximate approach increases as well.

Table 5.10 Comparison of the computation time ratio of exact (TE) and approximation (TA) approaches

Buffer size	30	40	45	50	55	60
TA/TE	0.02	0.01	0.005	0.004	0.003	0.001

5.4 State-Dependent Jump Priorities

In this section we consider models of switch with state-dependent jump priorities (JP). Specific property of these priorities consists in the following: in some situations low-priority cell (L-cell) might jump to queue of high-priority cells (H-cell) and thus become H-cell. The main questions arising at the introduction of the jump priorities are as follows: (1) determination of the instant of passage from the L-queue to the H-queue, (2) determination of the number of L-cells passing to the H-queue, and (3) determination of the state parameter (or parameters) which depend on the jump priorities.

Here we consider the jump priorities which are activated at the instants of arrival of the L-cells. This scheme for determination of JP is explained by the following reason: these priorities are introduced with the aim of increasing the chances of the L-cells to be serviced in an acceptable time, that is, to prevent their ageing in the queue, it is only natural to expect that they must be activated at the instants of their arrival.

Concerning the second problem, notice that only one L-cell could be passed in H-queue. This assumption is accepted to simplify the intermediate mathematical calculations, and therefore, it is possible to obtain analytically treatable results.

Finally, concerning the third problem, notice that here we investigate in detail only two kinds of jump priorities: (a) JP which depends on the number of L-cells in buffer and (b) JP which depends on the number of H-cells in buffer.

Note that here only models with finite (separate) buffers are considered since in practice switches have no infinite storage. At that, two kinds of jump priorities are investigated.

5.4.1 Various Schemes to Determination of State-Dependent Jump Priorities

First consider the model with JP which depends on the number of L-cells in buffer. Detailed description of the appropriate queueing model consists in the following.

Two Poisson flows of heterogeneous cells arrive to the input of the single-server system, the intensity of the ith flow being λ_i, $i = 1, 2$. The first flow is that of real-time cells (H-cells), whereas the second flow is that of the non-real-time cells (L-cells). The time of channel occupation is a random variable obeying the exponential distribution with the parameter μ for cells of both types. There are different buffers for the waiting heterogeneous cells, the size of the buffer for the cells of the ith type being R_i, $0<R_i<\infty$, $i=1,2$. We also notice that the H-cells have non-preemptive high priorities over the L-cells.

Note that the H-cells are always received with the probability one if at the instant of their arrival there is at least one free place in the H-buffer; otherwise, they are lost with the probability one. If at the instant of arrival of an L-cell the number of buffered cells of this type is k, $k<R_2$ and that there is a free place in the H-queue, then one L-cell immediately goes to the H-buffer with the probability $\alpha(k)$ (we assume for certainty that it is the L-cell at the queue head that goes to the H-buffer); the arriving L-cell is queued with the complementary probability $1-\alpha(k)$ if there is a free place on the queue. If there is no free place upon arrival of the L-cell on the H-queue, then with the probability one the arriving L-cell is added to the L-queue if it has a free place; otherwise, the L-cell is lost with the probability one.

If upon the arrival of an L-cell there is no free place on the L-queue, then the arriving L-cell is added to the H-queue with the probability $\alpha(R_2)$ if there is a free place. Otherwise, the arriving L-cell is lost with the probability $1-\alpha(R_2)$. We note that in the case of a successful jump the L-cell becomes an H-cell serviced and then as an H-cell according to the non-preemptive high-priority rules.

We notice that a number of the well-known servicing disciplines are obtained under particular values of the parameters $\alpha(k)$, $k=0,1,\ldots,R_2$. For example, for $\alpha(k)=0$ for all $k=0,1,\ldots,R_2$, the classical non-preemptive high-priority rule is obtained. Additionally, one can introduce a threshold scheme to determine the probabilities $\alpha(k)$, where $\alpha(k)=\alpha_i$ if $L_{i-1}\leq k<L_i, i=1,\ldots,r, L_0=0, L_0=R_2$. At that, the probabilities α_i, $i=1,\ldots,r$ can be determined using different techniques.

The state of buffers at an arbitrary time instant is describable by the two-dimensional vector $\boldsymbol{n}=(n_1,n_2)$, where n_i is the number of buffered i-cells, $i=1,2$. Stated differently, operation of this system follows the two-dimensional Markov chain with the state space:

$$S=\{\boldsymbol{n}:n_i=0,1,\ldots,R_i,\quad i=1,2\}. \tag{5.48}$$

The nonnegative elements of the Q-matrix of this 2-D MC (see Fig. 5.7) are given by

$$q(\boldsymbol{n},\boldsymbol{n}')=\begin{cases}\lambda_1+\lambda_2\alpha(n_2) & \text{if } \boldsymbol{n}'=\boldsymbol{n}+\boldsymbol{e}_1,\\ \lambda_2\delta(n_1,R_1) & \\ +\lambda_2(1-\alpha(n_2))(1-\delta(n_1,R_1)) & \text{if} \boldsymbol{n}'=\boldsymbol{n}+\boldsymbol{e}_2,\\ \mu & \text{if} n_1>0, \boldsymbol{n}'=\boldsymbol{n}-\boldsymbol{e}_1 \text{or} n_1=0, \boldsymbol{n}'=\boldsymbol{n}-\boldsymbol{e}_2,\\ 0 & \text{inothercases.}\end{cases} \tag{5.49}$$

All states of this 2-D MC are communicating, and, consequently, it is ergodic. The stationary state probability $\boldsymbol{n}\in S$ is denoted by $p(\boldsymbol{n})$. Solution of the

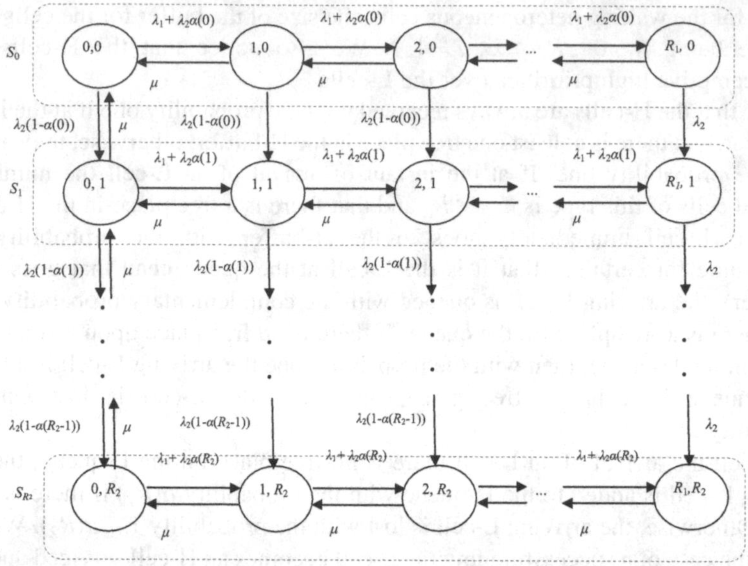

Fig. 5.7 State diagram of the model with jump priorities which depends on the number of L-cells in buffer

corresponding SGBE is the standard way of determining the stationary probabilities. It is constructed with regard for Eq. (5.49) and we left it to the reader.

After determining the state probabilities, one can establish QoS metrics of the switch. For example, the H-cell loss probability is given by

$$\mathrm{CLP}_1 = \sum_{k=0}^{R_2} p(R_1, k). \tag{5.50}$$

The L-cell loss probability is given by

$$\mathrm{CLP}_2 = \sum_{k=0}^{R_1-1} p(k, R_2)(1 - \alpha(R_2)) + p(R_1, R_2). \tag{5.51}$$

To determine the mean number of the heterogeneous cells in the queue the standard technique for determining the mean value of the discrete random variable is used (see Eq. (5.6)). The mean cell transmission delay for the heterogeneous cells is determined from Eq. (5.5).

Therefore, to determine the exact values of the QoS metrics one has to solve the SGBE for the state probabilities $p(\boldsymbol{n})$, $\boldsymbol{n} \in S$. It has no analytical solution, but the numerical methods of the linear algebra may be used to calculate these metrics. This approach (exact method) can be used only for the models of low dimensions and becomes inefficient with growing their dimensions. Therefore, a need arises for a more efficient approximate method to solve this problem. It has high accuracy for models in which intensity of H-cells is much more than intensity of L-cells, that is, it is assumed below that $\lambda_1 >> \lambda_2$. Note that this assumption is not extraordinary because introduction of the jump priorities for the L-cells makes sense, namely, in the systems with heavy intensity of H-cells.

Consideration is given to the following splitting of the state space of the model:

$$S = \cup_{i=0}^{R_2} S_i,\ S_i \cap S_j = \varnothing \quad \text{if } i \neq j, \tag{5.52}$$

where $S_i = \{\boldsymbol{n} \in S : n_2 = i\}$, $i = 0, 1, \ldots, R_2$.

The classes of microstates S_i in splitting Eq. (5.52) are united into individual merged states $\langle i \rangle$, and in the original state space (5.48) appropriate merge function is constructed. The stationary probability of the state (*k*, *i*) in the split model with the state space S_i is denoted by $\rho_i(k)$, $i = 0, 1, \ldots, R_2$, $k = 0, 1, \ldots, R_1$. Each split model with state space S_i is a 1-D BDP with the parameters (see Fig. 5.7):

$$q_i(k_1, k_2) = \begin{cases} \lambda_1 + \lambda_2 \alpha(i) & \text{if } k_2 = k_1 + 1, \\ \mu & \text{if } k_2 = k_1 - 1, \\ 0 & \text{in other cases.} \end{cases}$$

As can be seen from the last formula, to determine the desired $\rho_i(k)$, one can use the formulas for calculation of the state probabilities of the queueing system with state-independent intensity of the input traffic $M(\lambda_1 + \lambda_2 \alpha(i)) | M(\mu) | 1 | R_1$. A modified Kendall notation where the symbol M is followed by the parameters of the corresponding distributions is used here and below to denote the queueing system. Consequently,

$$\rho_i(k) = \theta_i^k \frac{1 - \theta_i}{1 - \theta_i^{R_1 + 1}}, \quad k = 0, 1, .., R_1,$$

where $\theta_i = v_1 + v_2 \alpha(i)$. The formulas for the case of $\theta_i \neq 1$ are given for brevity.

The elements of the generating matrix of the merged model (see Fig. 5.7) are given by

$$q(\langle i \rangle, \langle j \rangle) = \begin{cases} \lambda_2 (1 - \alpha(i))(1 - \rho_i(R_1)) + \lambda_2 \rho_i(R_1) & \text{if } j = i + 1, \\ \mu \rho_i(0) & \text{if } j = i - 1, \\ 0 & \text{in other cases.} \end{cases}$$

So, the probabilities of the merged states $\pi(\langle i \rangle)$, $\langle i \rangle \in \Omega$ are

$$\pi(\langle i\rangle) = \prod_{j=1}^{i} A_j\pi(\langle 0\rangle), \quad i = 1,2,\ldots,R_2,$$

where

$$A_j = v_2\frac{(1-\alpha(j-1))\big(1-\rho_{j-1}(R_1)\big)+\rho_{j-1}(R_1)}{\rho_j(0)}, \ \pi(\langle 0\rangle)$$
$$= \left(1+\sum_{k=1}^{R_2}\prod_{i=1}^{k} A_i\right)^{-1}.$$

After certain transformation, we establish that

$$\mathrm{CLP}_1 \approx \sum_{k=0}^{R_2}\rho_k(R_1)\pi(\langle k\rangle) \tag{5.53}$$

$$\mathrm{CLP}_2 \approx \pi(\langle R_2\rangle)(1-\alpha(R_2))\big(1-\rho_{R_2}(R_1)\big)+\rho_{R_2}(R_1) \tag{5.54}$$

$$Q_1 \approx \sum_{k=1}^{R_1}k\sum_{i=0}^{R_2}\rho_i(k)\pi(\langle i\rangle) \tag{5.55}$$

$$Q_2 \approx \sum_{k=1}^{R_2}k\pi(\langle k\rangle) \tag{5.56}$$

The parameters CTD_k, $k = 1,2$ are determined from Eq. (5.5).

Now consider the model with JP which depends on number of H-cells in buffer. As above, H-cell is accepted if upon its arrival there is at least one free place in corresponding buffer; otherwise, arrived H-cell is lost with probability 1. However, in this model if upon arrival of an L-cell the number of buffered H-cells is k, $k < R_1$, then one L-cell immediately goes to the H-buffer with the probability $\alpha(k)$; the arriving L-cell is queued in L-buffer with the complementary probability $1-\alpha(k)$ if it has a free place. If there is no free place upon arrival of the L-cell on the H-queue, then with the probability $1-\alpha(R_1)$ the arriving L-cell is added to the L-buffer if it has a free place; otherwise, the L-cell is lost with the probability $\alpha(R_1)$. If upon arrival of the L-cell both buffers are full, then arrived cell is lost with probability 1.

The two-dimensional state vector and state space for this model are determined by Eq. (5.48). However, here nonnegative elements of Q-matrix are calculated as follows (see Fig. 5.8):

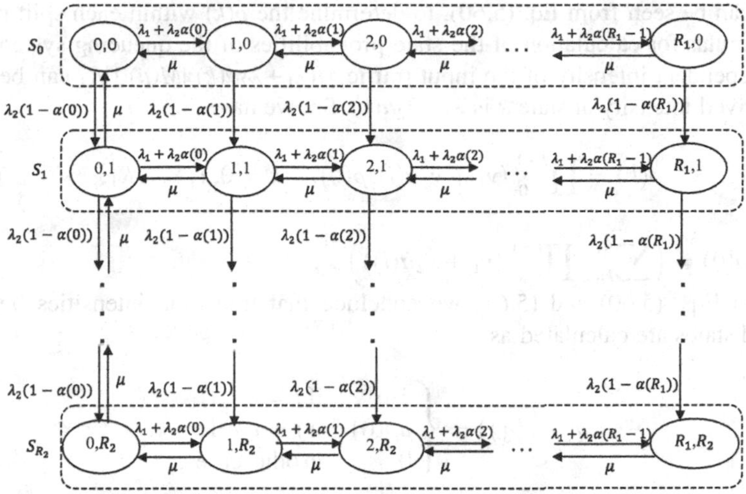

Fig. 5.8 State diagram of the model with jump priorities which depends on the number of H-cells in buffer

$$q(\boldsymbol{n},\boldsymbol{n}') = \begin{cases} \lambda_1 + \lambda_2\alpha(n_1) & \text{if } \boldsymbol{n}' = \boldsymbol{n} + \boldsymbol{e}_1, \\ \lambda_2(1-\alpha(n_1)) & \text{if } \boldsymbol{n}' = \boldsymbol{n} + \boldsymbol{e}_2, \\ \mu & \text{if } n_1 > 0,\ \boldsymbol{n}' = \boldsymbol{n} - \boldsymbol{e}_1 \text{ or } n_1 = 0,\ \boldsymbol{n}' = \boldsymbol{n} - \boldsymbol{e}_2, \\ 0 & \text{in other cases.} \end{cases} \tag{5.57}$$

Cell loss probabilities now are determined from following relations:

$$\text{CLP}_1 = \sum_{k=0}^{R_2} p(R_1,k); \tag{5.58}$$

$$\text{CLP}_2 = \sum_{k=0}^{R_1-1} p(k,R_2)(1-\alpha(k)) + \alpha(R_1)\sum_{k=0}^{R_2-1} p(R_1,k) + p(R_1,R_2). \tag{5.59}$$

Not repeating the stages of the approximate approach for solution to a similar problem, final formulas to calculate the QoS metrics are given below. In this case stationary distribution of the split models with state space S_i, $i = 0, 1, \ldots, R_2$, does not depend on index i, i.e., for any i, $i = 0, 1, \ldots, R_2$ transition intensities within split models with state space S_i are defined as (see Fig. 5.8)

$$q(k_1,k_2) = \begin{cases} \lambda_1 + \lambda_2\alpha(k_1) & \text{if } k_2 = k_1 + 1, \\ \mu & \text{if } k_2 = k_1 - 1, \\ 0 & \text{in other cases.} \end{cases} \tag{5.60}$$

As can be seen from Eq. (5.60), to determine the $\rho(k)$ within each split model, the formulas for calculation of the state probabilities of the queueing system with state-dependent intensity of the input traffic $M(\lambda_1+\lambda_2\alpha(k))|M(\mu)|1|R_1$ can be used, i.e., arrived intensity at state k is $\lambda_1+\lambda_2\alpha(k)$. So, we have

$$\rho(k) = \prod\nolimits_{i=0}^{k-1} (v_1 + v_2\alpha(i))\rho(0), \quad k = 0, 1, \ldots, R_1, \tag{5.61}$$

where $\rho(0) = \left(\sum\nolimits_{k=0}^{R_1} \prod\nolimits_{i=0}^{k-1} (v_1 + v_2\alpha(i))\right)^{-1}$.

From Eqs. (5.60) and (5.61) we conclude that transition intensities between merged states are calculated as

$$q(\langle i\rangle, \langle j\rangle) = \begin{cases} \lambda_2 a & \text{if } j = i+1, \\ \mu\rho(0) & \text{if } j = i-1, \\ 0 & \text{in other cases,} \end{cases}$$

where $a = \sum_{i=0}^{R1} \rho(i)(1-\alpha(i))$.

Thus stationary probabilities of merged states are

$$\pi(\langle k\rangle) = \theta^k \pi(\langle 0\rangle), \quad k = 0, 1, \ldots, R_2$$

where $\theta = v_2 a/(\rho(0))$.

Finally, approximate values of QoS metrics at using such kind of JP are calculated as follows:

$$\mathrm{CLP}_1 \approx \sum_{k=0}^{R_2} \rho(R_1)\pi(\langle k\rangle) = \rho(R_1); \tag{5.62}$$

$$\mathrm{CLP}_2 \approx \pi(\langle R_2\rangle)\left[\rho(R_1) + \sum_{k=0}^{R_1-1} \rho(k)(1-\alpha(k))\right] + (1-\pi(\langle R_2\rangle))\alpha(R_1)\rho(R_1); \tag{5.63}$$

$$Q_1 \approx \sum_{k=1}^{R_1} k\rho(k). \tag{5.64}$$

The mean number of the cells of type 2 is determined similar to Eq. (5.56).

Note that models with common buffers for heterogeneous cells and JP which depends on the total number of heterogeneous cells in buffers might be investigated by similar way.

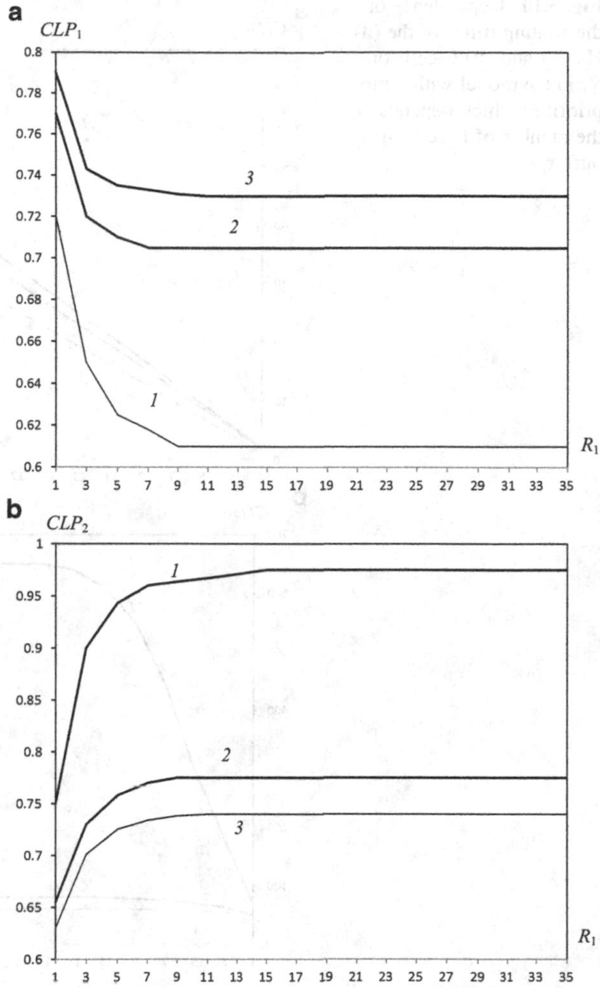

Fig. 5.9 Dependence of the loss probabilities of the (**a**) H-cells and (**b**) L-cells on R_1 in the model with jump priorities which depends on the number of L-cells in buffer

5.4.2 Numerical Results

Consider the results of numerical experiments obtained for the models with jump priorities. First consider the results for the model with JP which depends on the number of L-cells in buffer.

Here we present only a small part of the computational experiments for the sample model with the parameters $R_2 = 15, \lambda_1 = 2, \lambda_2 = 1$ and $\mu = 0.8$. Consideration is given to three schemes for determination of the jump priorities: (1) $\alpha(i) = 0$ for all $i = 0, 1, \ldots, R_2$, (2) $\alpha(i) = 0.7$ for all $i = 0, 1, \ldots, R_2$, and (3) $\alpha(i) = (i+1)/(i+2)$ for $i = 0, 1, \ldots, R_2$.

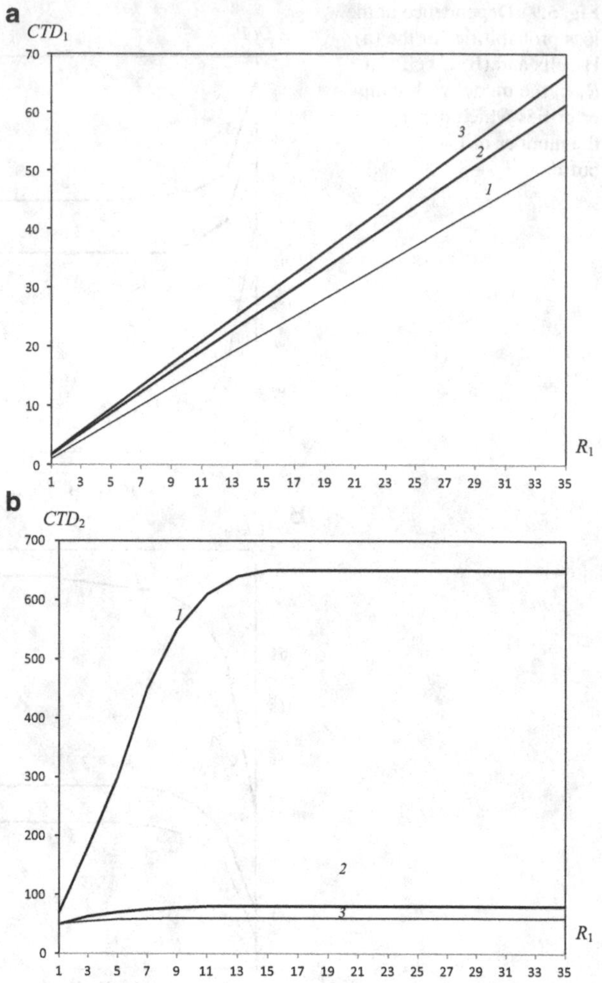

Fig. 5.10 Dependence of the waiting times of the (**a**) H-cells and (**b**) L-cells on R_1 in the model with jump priorities which depends on the number of L-cells in buffer

The results are shown in Figs. 5.9 and 5.10 where the numbers of the graphs 1, 2, and 3 correspond to schemes (1), (2), and (3).

The probability of losing the H-cells decreases with an increase of the buffer space for the cells of the given type, which is a quite expectable result (see Fig. 5.9a).

It deserves noting the identity in the nature of variations of the function CLP_1 under all schemes of determination of the parameters $\alpha(i)$, that is, for $R_1 > 5$ this function falls at a very low rate, which is due to the fact that the chances for arrival of L-cells in a buffer increase concurrently with an increase in the size of the buffer for H-cells. We note that in the case of using scheme (3) for determination of the parameters (αi), the value of the function CLP_1 is always greater than the corresponding values in the cases of using other schemes. It also deserves noting that scheme (2) stands in between the other two schemes.

An increase in the loss probability of the L-cells relative to the size of the buffer for H-cells also is an expectable result (sec Fig. 5.9b). It also deserves noting the identity in the nature of variations of the function CLP_2 under all schemes for determination of the parameters $\alpha(i)$, that is, for $R_1 > 5$, this function grows at a very low rate. In contrast to the function CLP_1, for the function CLP_2, scheme (3) is most desirable. These facts have a quite logical substantiation: the increasing function $\alpha(i)$ improves the chances for servicing of the L-cells. Scheme (2) here also is intermediate between the two other schemes.

The functions CTD_1 and CTD_2 are no decreasing relative to the size of the H-cell buffer, but the nature of their variations is different (see Fig. 5.10). For example, for all three schemes for determination of the parameters $\alpha(i)$, the function CTD_1 is almost linear (see Fig. 5.10a). The function CTD_2 is nonlinear only for small values of R_1, and for $R_1 > 5$, it becomes almost constant for schemes (2) and (3) (see Fig. 5.10b) and grows catastrophically for scheme (1). It also deserves noting that scheme (1) is the best one for the function CTD_1, and on the contrary, for the function CTD_2 it is scheme (3) that is the best one. For both functions, scheme (2) is intermediate between schemes (1) and (3).

Now consider the results of numerical experiments obtained for the model with JP which depends on the number of H-cells in buffer.

Below we present the results of computational experiments for the sample model with the parameters $R_2 = 20, \lambda_1 = 5, \lambda_2 = 2$ and $\mu = 10$. To define the jump priorities, two schemes are used: (1) $\alpha(i) = (i+1)/(i+2)$ for all $i = 0, 1, \ldots, R_1$ and (2) $\alpha(i) = 1/(i+2)$ for all $i = 0, 1, \ldots, R_1$. In other words, in scheme 1, parameters $\alpha(i)$ are increasing one's subject to i, $i = 0, 1, \ldots, R_1$, while in scheme 2 we have the inverse situation.

The results are shown in Figs. 5.11 and 5.12 where the numbers of the graphs 1 and 2 correspond to schemes (1) and (2), respectively.

The loss probabilities of the both kinds of cells decrease with an increase of the size of H-buffer (i.e., waiting space for the H-cells) in each scheme which are quite expectable results (sec Fig. 5.11). The identity in the nature of variations of the both loss probabilities under both schemes of determination of the parameters $\alpha(i)$ is observed, i.e., both functions CLP_k, $k = 1, 2$ are almost linear decreasing functions with respect to R_1 and their rate of change is high enough. The analysis of the graphs shows that intervals of changes of loss probabilities of heterogeneous cells in scheme 1 are quite close each other. However, in scheme 2, intervals of changes of loss probabilities of heterogeneous cells significantly differ from each other; it is accurately visible at great values of parameter R_1. Moreover, in scheme 2, the loss probabilities of L-cells become even less than the loss probabilities of H-cells. These facts show that by using the appropriate scheme to determination of jump probabilities, it is possible to control the loss probabilities of heterogeneous cells.

Remarkably that for the indicated QoS metrics CLP_k, $k = 1$, 2, scheme 2 is favorable. In other words, in scheme 2, values of loss probabilities of both kinds of cells are significantly less than in scheme 1. It means that the decreasing schemes

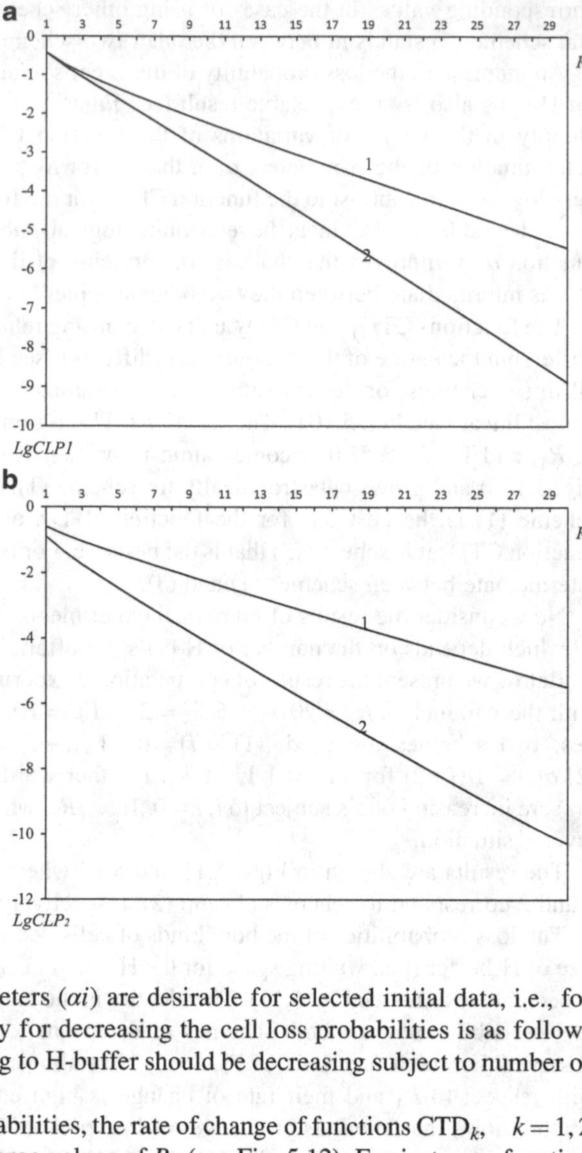

Fig. 5.11 Dependence of the loss probabilities of the (**a**) H-cells and (**b**) L-cells on R_1 in model with jump priorities which depends on the number of H-cells in buffer

for determination of parameters (αi) are desirable for selected initial data, i.e., for the given initial data, policy for decreasing the cell loss probabilities is as following: probabilities of jumping to H-buffer should be decreasing subject to number of H-cells in buffer.

Unlike the cell loss probabilities, the rate of change of functions $\text{CTD}_k,\quad k = 1, 2$ is very low especially for large values of R_1 (see Fig. 5.12). For instance, function CTD_2 is almost constant in scheme 1. Note that in scheme 1 the average cell transfer delay for H-cells is almost three times more than the appropriate metric for L-cells when $R_1 > 7$, while the difference between their values is about 15 % in scheme 2.

Remarkably unlike the cell loss probabilities, scheme 2 is favorable for H-cells, while for L-cells scheme 1 is favorable. Let's notice that when $R_1 > 10$, the values of function CTD_1 in scheme 1 are almost for 50 % more than its values in scheme 2, and values of function CTD_2 in scheme 2 are almost in two times more than its values in scheme 1.

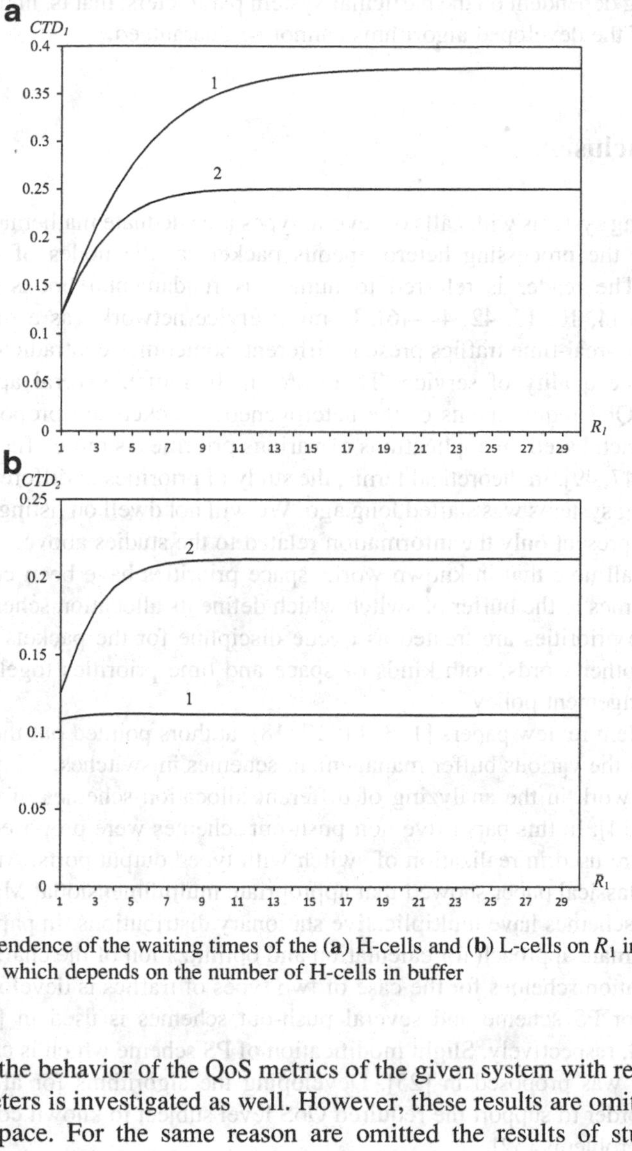

Fig. 5.12 Dependence of the waiting times of the (**a**) H-cells and (**b**) L-cells on R_1 in model with jump priorities which depends on the number of H-cells in buffer

Note that the behavior of the QoS metrics of the given system with respect to its other parameters is investigated as well. However, these results are omitted for the reasons of space. For the same reason are omitted the results of studying the accuracy of the developed approximate formulas. At that, the exact values of the QoS metrics for the models of moderate size were established from the corresponding SGBE (some analytical considerations about the high accuracy of these formulas can be found above). Here we just note that under the above assumption about the system loads, the exact and approximate values of the QoS metrics differ in the worst case only at the third position after decimal point. At the same time, rejection of this assumption leads to substantial errors, their absolute values being dependent on the particular system parameters, that is, in this case high accuracy of the developed algorithms cannot be guaranteed.

5.5 Conclusion

The queueing systems with calls of several types are adequate mathematical models to describe the processing heterogeneous packets at the nodes of multiservice networks. The reader is referred to numerous fundamental books for detailed information [4, 10, 13, 42, 44–46]. In multiservice networks, as a rule, the real-time and non-real-time traffics present different, sometimes contradictory, requirements on the quality of service. Therefore, in literature, several approaches to satisfy the QoS requirements of the heterogeneous packets are proposed. Among them approach based on applications of various priorities is more efficient [12, 15, 16, 41, 43, 47, 49]. In theoretical terms, the study of priorities in different models of the queueing systems was started long ago. We will not dwell on listing the existing results and present only the information related to the studies above.

First of all note that in known works space priorities have been considered as access schemes to the buffer of switch which define its allocation schemes; respectively, time priorities are treated as queue discipline for the packets in buffer of switch. In other words, both kinds of space and time priorities together define a buffer management policy.

In excellent review papers [1, 3, 11, 17, 18], authors pointed out the major pros and cons of the various buffer management schemes in switches.

Pioneer work in the analyzing of different allocation schemes of shared finite storage is [14]. In this paper five non-push-out schemes were proposed and nowadays they are used in realization of switch with typed output ports. Authors of the indicated classical paper showed that appropriate multidimensional Markov chains for all five schemes have multiplicative stationary distributions. In papers [26–29], the approximate approach for calculation and optimization of the characteristics of these allocation schemes for the case of two types of traffics is developed. Similar approach for PS scheme and several push-out schemes is used in [30, 31] and [32, 35, 36], respectively. Slight modification of PS scheme which is called limited PS scheme was proposed in [25]. Developing the algorithms for allocation, the buffers in order to support the required QoS level subject to known constraints are important problems [52].

Multidimensional Markov chains which are models of switches with push-out schemes for use in a shared finite buffer storage have no multiplicative stationary distributions. Therefore their analysis becomes very difficult. Such models are investigated in many works via simulation techniques (see, e.g., [5, 6]). The papers [7, 9, 48, 50, 51, 54] are devoted to finding a form of optimal access scheme in the class of PO schemes.

Note that a new type of the so-called multiple priorities was thoroughly studied during the last decade [2, 8, 19, 33, 37]. In distinction to the classical priorities, in the case of multiple priorities, the real-time cells have high time priorities and low space priorities, and the non-real-time cells, low time priorities and high space priorities. In this section a simple numerical approximation method to calculate the QoS measures of the queueing model with multiple space and time priorities is proposed. Space priorities are realized by the non-push-out and push-out schemes, whereas time priorities are realized by fixed non-preemptive priority scheme. This approach allows developing a simple computational procedure to find the desired QoS metrics of the model and solve the problem of finding the appropriate values of the proposed access scheme's parameter under restrictions to both the loss probability of the loss-sensitive traffic and the waiting time of the delay-sensitive traffic.

The recent publications study a new type of priorities—jump priorities. The first paper in this area was [20] whose authors investigate the time-dependent jump priorities. In this model for each type of calls, the deterministic type-dependent parameters are defined, and if the waiting time of an i-call at the head of the ith queue reaches the same threshold, then the call goes to the queue $i-1, i=2,\ldots,N$, and this process goes on until a call of any type gets access to a channel or reaches a queue with the highest priority, that is, the queue number one. Formulas for calculation of the mean waiting time of the heterogeneous calls were established in [20]. Note that the jump priorities proposed in [20] are inconvenient for realization in switches because they need additional facilities to monitor the waiting time for the heterogeneous cells.

For the discrete-time queueing system, different kinds of JP were proposed in [21–24, 53]. In these papers, models with infinite buffers for waiting calls of each type are examined. A scheme, of head-of-line merge-by-probability (HOL-MBP) according to which at the end of each time slot all L-calls go to the end of the queue of H-calls with the probability $\beta, 0<\beta<1$, was proposed in [21]. A modification of the HOL-MBP scheme was considered in [22]. It was named head-of-line jump-or-serve (HOL-JOS), and, in contrast to the scheme of [21], in it only one L-call goes from the queue head into the H-queue. In contrast to the HOL-JOS scheme, in the head-of-line jump-if-arrival (HOL-JIA) scheme [24], the transition of the L-call into the H-queue depends not only on the state of the H-queue at the beginning of the slot but also on the number of arrivals of L-calls during this slot. The only distinction of the first HOL-JIA scheme [24] from the second HOL-JIA scheme [23] lies in that in the latter scheme the L-calls can pass immediately to the H-queue.

Formulas for the generating functions of the call queue lengths of both types and the time of H-call waiting on the queue, as well as their moments, were developed in [21–24, 53]. Additionally, the mean time of waiting on the queue of L-calls was determined.

It deserves noting that [21–24, 53] are devoted to studying the models of queueing systems with infinite buffers which cannot be used as adequate models of switches in packet switching networks because, as a rule, their switches have limited buffers for temporal storage of the heterogeneous cells. Stated differently, for wide introduction of the jump priorities, their efficiency in the PSN must be determined. So, the papers [34, 38–40] introduce for the continuous-time queueing systems a new class of randomized jump priorities. They make it possible to pass from the L-queue into the H-queue only at the instants of arrival of the L-calls, the probability of such transitions depending on the number of heterogeneous calls in the system. Introduction of the constraints on the size of buffer (buffers, in the case of queueing systems with separate queues) for waiting for the heterogeneous cells necessitates the determination of a new QoS metrics, the cell loss probability. Another distinction of the papers [34, 38–40] from [21–24, 53] lies in using an approach based on the space merging theory of the states of the 2-D MC for model analysis. This approach enabled development of simple computational procedures for determination of all QoS metrics of the switches under consideration.

Note that it is possible to solve various optimization problems of the switch with jump priorities. For example, of great scientific and practical interest is the problem of determining the optimal threshold scheme under the given constraints on all (or some) system QoS metrics. Additionally, the problem of determining the optimal values of the introduced probabilities of transition of the L-calls to the H-queue where these probabilities are controllable also deserves consideration. For the last problem, solution may be based on the methods of the theory of Markov decision processes, and various system QoS metrics may be used as the optimization criteria; consideration can be given also to the multi-criteria problems. The search of the optimal buffer sizes for maintaining the given level of the quality of service is also of interest.

References

1. Arpachi M, Copeland JA (2000) Buffer management for shared-memory ATM switch. IEEE Commun Tutorials 3(1):2–10
2. Chao HJ, Uzun N (1995) Queue management with multiple delay and loss priorities for ATM switches. IEEE/ACM Trans Netw 3(6):652–659
3. Chatranon G, Labrador MA, Banerjee S (2004) A survey of TCP friendly router-based AQM schemes. Comput Commun 27(15):1424–1440
4. Chen TM, Stephen SL (1995) ATM switching systems. Artech House, Boston, MA
5. Choudhury AK, Hahne EL (1997) A new buffer management scheme for hierarchical shared-memory switches. IEEE/ACM Trans Netw 5(5):728–738
6. Choudhury AK, Hahne EL (1998) Dynamic queue length thresholds for shared-memory packet switches. IEEE/ACM Trans Netw 6:130–140
7. Cidon I, Georgiadis L, Guerin R, Khamisy A (1995) Optimal buffer sharing. IEEE J Sel Area Commun 13(7):1229–1240
8. Demoor T, Fiems D, Walraevens J (2011) Partially shared buffers with full or mixed priority. J Ind Manag Optim 7(3):735–751
9. Foschini GJ, Gopinath B, Hayes JF (1983) Sharing memory optimally. IEEE Trans Commun 31:352–359
10. Gebali F (2008) Analysis of computer and communication networks. Springer, New York
11. Guerin R, Peris V (1999) Quality-of-service in packet networks: basic mechanisms and directions. Comput Netw 31:169–189
12. Huang TV, Wu JLC (1994) Performance analysis of ATM switches using priority schemes. IEE Proc Commun 141(4):248–254
13. Hui JY (1992) Switching and traffic theory for integrated broadband networks. Kluwer, Boston, MA

14. Kamoun F, Kleinrock L (1980) Analysis of shared finite storage in a computer network node environment under general traffic conditions. IEEE Trans Commun 28(7):992–1003
15. Koukos AK (2007) Queuing monitoring system for a multi-mode shared-buffer for IBCN networks. Int J Electron 94(11):1059–1074
16. Kroner H, Hebuterne G, Boyer P, Gravey A (1991) Priority management in ATM switching nodes. IEEE J Sel Area Commun 9(3):418–427
17. Kuehn PJ (1996) Reminder on queueing theory for ATM networks. Telecommun Syst 5 (1):1–24
18. Labrador MA, Banerjee S (1999) Packet dropping policies for ATM and IP networks. IEEE Commun Surv 2(3):2–14
19. Lee Y, Choi BD (2001) Queuing system with multiple delay and loss priorities for ATM networks. Inform Sci 138:7–29
20. Lim Y, Kobza JE (1990) Analysis of delay dependent priority discipline in an integrated multiclass traffic fast packet switch. IEEE Trans Commun 38(5):659–665
21. Maertens T, Walraevens J, Bruneel H (2006) On priority queues with priority jumps. Perform Eval 63(12):1235–1252
22. Maertens T, Walraevens J, Bruneel H (2007) A modified HOL priority scheduling discipline: performance analysis. Eur J Oper Res 180(3):1168–1185
23. Maertens T, Walraevens J, Bruneel H (2008) Performance comparison of several priority schemes with priority jumps. Ann Oper Res 162:109–125
24. Maertens T, Walraevens J, Moeneclaey M, Bruneel H (2006) A new dynamic priority scheme: performance analysis. In: Proceedings of 13th international conference on analytical and stochastic modeling techniques and applications, pp 74–84
25. Matsufuru N, Nishumira K, Aibara R (1998) Performance analysis of buffer management mechanisms with delay constraints in ATM switches. IEICE Trans Commun E81-B:431–439
26. Melikov AZ, Akperov VO (2003) Calculation and optimization of parameters of service quality in Ethernet networks. Automat Contr Comput Sci 37(6):55–63
27. Melikov AZ, Fattakhova MI (2002) Computational algorithms to optimization of buffer allocation strategies in a packet switching networks. Appl Comput Math 1(1):51–58
28. Melikov AZ, Fattakhova MI (2002) Approximate analysis of quality of service in integrated network nodes. Automat Contr Comput Sci 36(2):34–40
29. Melikov AZ, Fattakhova MI (2003) Problems of optimization of the indicators of service quality in the nodes of integrated networks. Automat Contr Comput Sci 37(1):56–61
30. Melikov AZ, Fattakhova MI, Feyziyev VS (2002) Analysis and optimization of the strategy of partial distribution of a buffer in ATM networks. Automat Contr Comput Sci 36(5):46–54
31. Melikov AZ, Fattakhova MI, Feyziyev VS (2004) Optimization of partial sharing strategy for buffer allocation subject to probabilities of packets blocking requirements. Automat Contr Comput Sci 38(1):37–43
32. Melikov AZ, Fattakhova MI, Nagiyev FN (2004) State space merging approach to optimization of push-out strategies in packet switching networks. Cybern Syst Anal 40(2):238–244
33. Melikov AZ, Feyziev VS, Rustamov AM (2006) Analysis of model of data packet processing in ATM networks with multiple space and time priorities. Automat Contr Comput Sci 40 (6):38–45
34. Melikov AZ, Oh Y, Kim CS (2013) A space merging approach to the analysis of the performance of queuing models with buffers and priority jumps. Ind Eng Manag Syst 12 (3):274–280
35. Melikov AZ, Ponomarenko LA, Fattakhova MI (2004) Numerical methods to studying of multi-flow queuing systems with virtual partition of common buffer. Cybern Syst Anal 40 (6):928–935
36. Melikov AZ, Ponomarenko LA, Fattakhova MI (2004) The push-out strategy with virtual threshold to access the integral network node buffer. J Autom Inform Sci 36(7):49–57
37. Melikov AZ, Ponomarenko LA, Kim CS (2007) Approximate method for performance analysis of queuing systems with multimedia traffics. Appl Comput Math 6(2):1–8

38. Melikov AZ, Ponomarenko LA, Kim CS (2012) Algorithmic approach to analysis of queuing system with finite queues and jump-like priorities. J Autom Inform Sci 44(12):43–53
39. Melikov AZ, Ponomarenko LA, Kim CS (2013) Approximate method for analysis of queuing models with jump priorities. Autom Rem Contr 74(1):62–75
40. Melikov AZ, Ponomarenko LA, Kim CS (2013) Numerical method for analysis of queuing models with priority jumps. Cybern Syst Anal 49(1):55–61
41. Mondragon R, Moore A, Pitts JM, Schormans JA (2009) Analysis, simulation and measurement in large-scale packet networks. IET Commun 3(6):887–905
42. Nader FM (2007) Computer and communications networks. Prentice Hall, Englewood Cliffs, NJ
43. Piet VM, Bart S, Guido HP (1997) Performance of cell loss probability management schemes in a single server queue. Int J Commun Syst 10(4):161–180
44. Pitts JM, Schormans JA (1997) Introduction to ATM design and performance. Wiley, Chichester
45. Saito H (1994) Teletraffic technologies in ATM networks. Artech House, Boston, MA
46. Schwartz M (1996) Broadband integrated networks. Prentice Hall, New York
47. Shan ZC, Liemin Y (1995) A new priority control of ATM output buffer. Telecommun Syst 4:61–69
48. Sharma S, Viniotis L (1995) Optimal buffer management policies for shared-buffer ATM switches. IEEE/ACM Trans Netw 7(4):575–587
49. Takahi Y, Hino S, Takahashi T (1991) Priority assignment control of ATM line buffers with multiple QoS classes. IEEE J Sel Area Commun 9(7):1078–1092
50. Tassilias L, Hung YC, Panwar SS (1994) Optimal buffer control during congestion in an ATM network node. IEEE/ACM Trans Netw 2(4):374–386
51. Thareja AK, Agrawala AK (1974) On the design of optimal policy for sharing finite buffers. IEEE Trans Commun 32:737–740
52. Tung TY, Chang JF (1997) Resource allocation algorithms for ATM nodes supporting heterogeneous traffic sources subject to varying quality of service requirements. IEICI Trans Commun E80-B:420–433
53. Walraevens J, Steyaert B, Bruneel H (2003) Performance analysis of single-server ATM queue with priority scheduling. Comput Oper Res 30(12):1807–1829
54. Wu GL, Mark JW (1995) A buffer allocation scheme for ATM networks: complete sharing based on virtual partition. IEEE/ACM Trans Netw 3:660–670

Index

A. Melikov, L. Ponomarenko, *Multidimensional Queueing Models in Telecommunication Networks*, DOI 10.1007/978-3-319-08669-9

Printed in the United States
By Bookmasters